ALSO BY CHRIS FERRIE

Where Did the Universe Come From? And Other Cosmic

Questions: Our Universe, from the Quantum to the Cosmos

The Cat in the Box

Quantum Physics for Babies and many other books

in the bestselling Baby University series

Quantum Bullsh*t

How to Ruin Your Life with Advice from Quantum Physics

Chris Ferrie

 sourcebooks

Published by Sourcebooks
P.O. Box 4410, Naperville, Illinois 60567-4410
(630) 961-3900
sourcebooks.com

Library of Congress Cataloging-in-Publication Data

Names: Ferrie, Chris, author.
Title: Quantum bullshit : how to ruin your life with advice from quantum
    physics / Chris Ferrie.
Description: Naperville, Illinois : Sourcebooks, [2023]
Identifiers: LCCN 2022006648 (print) | LCCN 2022006649
(ebook) | (trade paperback) | (epub)
Subjects: LCSH: Quantum theory--Popular works.
Classification: LCC QC174.123 .F48 2023 (print) | LCC QC174.123 (ebook) |
    DDC 530.12--dc23/eng20220726
LC record available at https://lccn.loc.gov/2022006648
LC ebook record available at https://lccn.loc.gov/2022006649

Printed and bound in the United States of America.
VP 10 9 8 7 6 5 4 3 2 1

To Albert Einstein

You would have hated this

What the fuck is this book?

Hi. I'm Chris. I'm the author of the bestselling children's science book *Quantum Physics for Babies*. If you are reading this aloud to a child right now, please stop. This book is not for children—well, at least not for children in 2023. Perhaps it is really 3023. In which case, go you, thirty-first-century hipster reading the classics. In 3023, maybe f-bombs are rated G, or maybe there is no rating system at all because it's a postapocalyptic world where children are badasses who fight robots and shit. But let's not start the first paragraph on a tangent.

So where was I? Oh, yes, I wrote this book. I suppose you want to know what it's all about. First, I'm going to assume you picked up this book for two reasons: (1) my name is on it and I'm awesome, and (2) you are utterly confused by quantum physics...and I'm still awesome.

Either you just want to get your schadenfreude tickled as we laugh along at the inanity with which our fellow humans have bastardized quantum physics, or you really want to know what makes quantum bullshit stink. By quantum bullshit, I mean the endless inane ways people stick the word *quantum* in front of whatever dubious product they're trying to sell you, hoping to convince you that their product is somehow both mysteriously mystical and scientifically proven. I've written this book as if you were of the latter camp, though I'm confident you'll still get a few chuckles along the way. By the time we are done, if you aren't laughing, you'll definitely be crying.

You've probably figured out already that the subtitle of the book is sarcasm; I don't actually want you to ruin your life. In fact, I want to save you from quantum bullshit. You clearly have some interest in me (who can blame you?) but also a deep fascination with quantum physics. I mean, I definitely can't blame you for that—quantum physics is pretty rad. But I went and wasted fifteen years of my life studying it. Uh, I mean, I totally didn't *waste* fifteen years of my life studying it. Shit. That sounds less convincing, doesn't it? Anyway, you wisely didn't do that. Your fascination stems from spectacular claims about quantum physics from science journalists and social media hot takes. Unfortunately, these have only served to confuse you further.

But it gets worse (for you, anyway). Not only is quantum physics confusing, it is also super important. It is the basis of all modern technology. So you should probably understand at least some of it, yeah? But how can something so important be so fucking difficult? I'm here to tell you it's not your fault you don't get it. Well, maybe it's a little bit your fault? It's certainly not my fault; at least we can agree on that. The situation is really unfortunate. Clearly, I'm not recommending you spend fifteen years studying quantum physics. But I do recommend you spend a few hours here with me, and I'll give you enough quantum knowledge to shield you from the worst of the bullshit spouted about it.

Here's the thing. Quantum physics is by far the most accurate scientific theory ever invented. It allows us to understand the structure of matter so we can build—atom by atom—materials that don't exist anywhere else in the universe. It allows us to understand what stars are made of and what lies beyond our telescopes deep in the cosmos. It allows us to build clocks that won't lose a second in the lifetime of the universe. It gave us lasers, medical scanners, and the computer you stole this book on the internet with. Quantum fucking physics. I'd marry it if I could.

Yet this isn't just another book about quantum physics. This book is about bullshit. But it's not about any old bullshit. This book is about quantum bullshit. Yeah. The good stuff. If you were a connoisseur of bullshit, you'd chef's kiss this crap. Quantum bullshit makes run-of-the-mill bullshit almost palatable. Don't worry. I'm not going to suggest you eat shit or anything. But I will tell you about some people that ought to.

For every scientist or engineer who has actual expertise in the topic, there is someone else out there who claims to understand quantum physics and says that knowledge has changed their life. Perhaps *using* quantum physics has healed some ailments, granted them financial success, or given them spiritual enlightenment. This is not just unfortunate—it's fucking bullshit. I'm going to tell you why.

Western culture has a love-hate relationship with science.

Many hate that it represents progress, and, fuck, who ordered that? I'm doing perfectly fine here watching reruns of Clint Eastwood movies next to my tire fire. At the same time, though, I'll happily head down to the hospital to get the latest scientific medical advice when my foot turns green from that buckshot my cousin accidentally hit me with. What's in this medicine? Science, you say? That sounds swell—load me up. People love science when it confirms their biases or saves their asses. And therein lies the problem.

There are some sly individuals out there—let's call them bullshitters—who want your money. I mean, I'll take your money if you are giving it away, but I'm not going to lie to get it. Not all bullshit contains blatant lies though; the "best" misrepresentations of science are subtle. There is just enough science-sounding fluff on the bullshit to make it sound credible. And what is the perfect esoteric science no one is even prepared to question? Quantum fucking physics.

I'm not sure if you are ready for this yet, but try this out. Type the word "quantum" into Google, followed by any noun you like. Unless the second word is "physics," you can pretty much bet that the results are mostly bullshit. Actually, wait, never mind. I searched #quantumphysics on Instagram, and it's *all* bullshit.

Quantum healing, quantum mysticism, quantum love, quantum crystals, quantum consciousness, quantum mediation, quantum energy...none of this has anything to do with quantum physics. But now we are at an impasse, for it seems one requires detailed knowledge of the subject to see why this is all crap. Until now.

In these pages, you will come to understand enough quantum physics to shield yourself from bullshit. But we are going to do things a little differently from most books on quantum physics. I'm *not* going to tell you that it's weird and mysterious and it's going to shatter your conception of the world. I'm not even going to tell you what quantum physics *is*. I'm going to tell you what it *isn't*.

Okay, maybe I'll tell you a little bit about what it is.

The truth is that quantum physics *is* difficult—as difficult as anything else people get doctoral degrees in. Let's take economics, for example. You might recall a recent experience of seeing some economist talking on the news. Somehow, when you heard their words, you felt like you had understood something. But the reality is that an economist's job is just as difficult as a quantum physicist's—performing detailed calculations using advanced mathematics. It's just that when the economist talks, she starts with things relevant to you, like money. When a quantum physicist talks...well, I don't know. They don't let quantum physicists speak on the news.

Economics is about goods and services and money. Even if these things actually have complicated technical meanings in an advanced field of specialized knowledge, you still *feel* some connection to them. Quantum physics, on the other hand, is about a part of the world we do not have direct access to. No one has ever *seen* an atom with their eyes. An atom is a name we give to a thing that appears in our *theories*. You want to imagine it's a *real* thing, but it isn't. It's an idea. We can only point to it as a symbol in long equations. By now, you're thinking, *Yawn! You are already boring me. Can't you just repeat some vague banalities about real estate prices so I have something to say to my neighbor during our morning small talk?*

So what went wrong? It's complicated. On one hand, the objects of study in quantum physics are far removed from our everyday experiences—we can't even see them! On the other, we—the quantum physicists—still haven't figured out how to talk about the stuff in plain language to our own colleagues, let alone an outsider. But we went ahead and started talking about quantum physics in public anyway. And having actual living scientists telling enthusiasts that quantum physics is magical and mysterious is how we get things like quantum healing jewelry, which—if you haven't seen where this is going by now—is bullshit.

I always found the mystery surrounding quantum physics quite ironic. Scientific communicators love to retell the story of the scientific revolution as being one of the triumphs of science over superstition. One devotes a life to science in an effort to dispel mystery, not create it. Such a task is arduous. After all, humans have a fascination with the mysterious. But science is what allows us today to read fictitious books and watch fictitious movies knowing full well what is depicted is fantasy. Great. I love to suspend disbelief while I watch superheroes break the laws of physics, knowing the world will still be intact when I leave the theater. But then physicists had to go and ruin everything by promoting the "mystery" of the quantum.

Take a look at the following table, which compares the top books recommended to me in quantum physics and economics. In all cases, these books are written by experts. Notice the difference?

Quantum Physics Titles	Economics Titles
The Quantum Astrologer's Handbook	The Wealth of Nations
Through Two Doors at Once	Capital in the Twenty-First Century
Beyond Weird: Why Everything You Thought You Knew about Quantum Physics Is Different	Thinking, Fast and Slow
Quantum Enigma	Nudge: Improving Decisions about Health, Wealth, and Happiness
Reality Is Not What It Seems	Freakonomics[1]

I've read at least one of these quantum physics books, and it wasn't all bad. The titles, however, are very revealing. Why the hell would you attempt to educate someone by telling them upfront they are going to be confused by it? The short answer is simple: this is the way it has always been done (and some of the people doing it were given Nobel Prizes) so why fix what isn't broken, right? Right? Wrong.

So now we are stuck with this perception, endorsed by experts, that quantum physics is mysterious. Hmm...wait a minute...you know what else is mysterious? Love, success, luck, health, who will win the Superbowl. Enter the charlatans. It's an

---

1 Okay, okay, you can't win every time.

easy argument—love is mysterious, and so is quantum physics; therefore, they are the same.

You know what? I already have a good feeling about you. So I'm going to tell you a secret. You can't tell anyone or use this to dupe foolish people who haven't read my book. With great power comes great responsibility, or some shit. So here it is, the four steps to quantum success in selling your crap.

1. State that your shitty product will solve some problems faster and more effectively than solutions provided by professionals like, for example, real doctors.
2. Acknowledge the seemingly miraculous nature of this claim as if it is a mystery to be solved.
3. State that the mechanism is complicated but works on the principles of quantum physics.
4. Say that this solved the aforementioned mystery, resolving all conflicts, and hence pacifying your victim...err... customer.

Now, how about that back pain? I have this rock...I mean, I have this crystal that will clear it right up for you.

You didn't realize that was sarcasm? Dear lord, we have a long way to go.

1

Quantum fucking energy

Energy. It's all around us. It's inside us. It's the life force of the universe, and it binds us to the cosmos. We are woven into the fabric of space-time with quantum threads. Whoa. That's fucking deep. Or is it just bullshit? If you are into this kind of crap, I'm sorry to burst your bubble. But it is bullshit. For every scientist who uses the word *energy* in a legitimately useful way, a charlatan uses it to take your money.

Energy is the most abused concept in science. Apparently, it's in healing crystals, you can balance it, you can "tap into it," and you can even use it to telekinetically arm yourself with a lightsaber. Okay, the last one is pretty cool—but still, bullshit. And when you ask what exactly this energy is inside the healing crystal, your guru might give you some profound sounding quantum fuckery. It's. All. Bullshit.

But quantum energy *is* real. It's just not what you think it is. That is, energy is not what you think if you get your information from wellness influencer memes shared by people you went to high school with and have never spoken to since but are friends with on Facebook anyway. No, sorry, what's-your-face, I don't want to join your wellness product pyramid scheme. Remember when Facebook was just that site where you "poked" people and waited a week for them to reply? Ah, the good old days...

Daddy, Where Does Energy Come From?

Energy has been around since forever—literally and figuratively! The universe started as a speck of energy and then—BANG!—matter and stuff. Or so we think. The Big Bang theory is *just* a theory, as they say. But it's a pretty damn good theory and currently the best one we've got. A new theory will eventually surpass it, but this is science, kids. We go with what works, what agrees with our observations and is helpful...until something better comes along. That something better will not be some magical theory combining love, physics, and the eternity of one's soul. It's going to be a well-thought-out mathematical model. No single theory is the capital-T Truth, but that doesn't mean all theories are equal. The one you heard about at the dog park, for example, is probably a complete crock. But I digress. Where was I? Oh yes, the point is energy, the physical stuff that gives rise to the changing motion of objects, has been here since long before us humans.

That was energy, *the stuff*. Energy, *the idea*, as invented by humans, has only been around a couple of millennia, basically since people started smoking out of pipes and writing down their thoughts. But only a few hundred years ago did the idea of energy start to take physical shape. In particular, the first technical definition in all its fancy-language glory was the following:

The product of the mass of a body and the square of its speed may properly be termed its energy.

That is, energy = mass × speed × speed. Speed is so nice, we multiply it twice. Since speed is distance divided by time, energy is mass times distance squared divided by time squared. Geez, that's a mouthful. Anyway, this definition also gave energy *units*. One unit of energy, now called a joule, is equal to one kilogram-meter squared per second squared. Take a large apple, hold it a foot above the ground. If you want to get more precise, pick out a 100-gram apple and hold it one meter above the ground.[2] It took you one small joule of energy to get it there, and it will release the same amount of energy when it falls. The moment before it hits the ground, it will be traveling sixteen kilometers per hour. Energy. It's measurable. It's precise. An ancient life force permeating the universe, on the other hand, it is not.

The idea of a supernatural energy source survives today and has evolved into many forms over the centuries. This goes to show how attractive the idea is. Nevertheless, whatever you want to call it, any prescientific notion of energy is just that: prescientific. In fact, it is demonstrably *nonscientific*, meaning it can be shown to have no basis in reality. It's kind of like reality TV— it's real in the sense that it is really a TV show with real people pretending to do real things, but it's not *reality*. There are plenty of *concepts* of energy but only one reality.

That's all fine and good if you don't know any better. But of course, today, we do know better thanks to science. But that

---

2 Yeah, we're going metric. Suck it, imperialists!

same science—the one that gives us measurable results—is being used against us! Today, peddlers of New Age wisdom use scientific jargon to defend their ideas. In the early days, the motion of the stars and planets was the false cause of nonsense prophecy. During the scientific revolution, magnetism or gravity provided the mechanism for alternative medical treatments or free energy. But the only game in town now is quantum, baby!

Never Make Predictions If You Plan on Becoming Famous

Before I tell you *what* quantum energy is, we need a slight digression on *why* quantum energy is. Imagine you are "notable" enough to have your own Wikipedia page. Now imagine you said something so profoundly foolish that there is an entire section of your Wiki biography titled "Pronouncements later proven to be false." Wow. You must have said some really stupid stuff. That or you are Lord Kelvin, famous nineteenth-century physicist, who said, "There is nothing new to be discovered in physics now. All that remains is more and more precise measurement."[3]

Lord Kelvin is so old that he might not have ever even ridden in a car. I think it is safe to say a little bit of physics was discovered after his time. In fact, there were plenty of observed phenomena without an explanation *during* his life. But let's not fault Kelvin too much. He was really old and cranky by then and probably too busy writing angry letters to people about the age of the Earth, which he also got wrong by a few billion years. If he were alive today, he'd be the kind of person TYPING

---

3 Okay, maybe he didn't say exactly this, but we'll let the historians figure it out. Lord knows they have the time.

INSULTING THINGS IN ALL CAPS ON TWITTER, which is not very becoming of a leader in science—or a country.

But now, two hundred years later, the world is in color, and we have quantum physics. Life is good. But really, that shit is complicated. Who the fuck ordered that? Well, let me tell you, it wasn't for fun. No one had fun before the 1980s. It was actually because a bunch of old European dudes couldn't figure out why stuff glows when it gets hot. Yeah. It's embarrassing.

When something gets really hot, it glows. Even a child knows this—after touching that irresistible glowing stovetop, that is. But here is the exciting part: the color with which the thing glows is the same *no matter what it is made of.* When a piece of iron is 1000°C, it glows red. When lava pouring out of a volcano is 1000°C, it glows red. When you set your oven to 1000°C...well, you'll probably start a fire, but not before that pot pie glows red hot!

This is a pattern. When an unexplained pattern is observed in nature, it gets scientists giddy with glee. It's like surprising a child with ice cream, except instead of a child, it's a grown-ass adult, and instead of ice cream, it's a table of numerical data. Okay, it's not like that at all.

Now, you may be thinking, *I've seen some elderly folk knitting patterns all day while watching soaps and twenty-four-hour news. What can possibly be so exciting about that?* I'm not talking about that kind of pattern. *Okay, well, the tile in my bathroom is patterned.* Yep. Now you are getting the *mathematicians* excited. But to get *physicists* really excited, you need to find a pattern in nature that is not obvious. Better yet is a pattern that cannot be explained with current theories. *Why the hell would you enjoy not having an explanation for something,* you might be thinking. What can I say? Physicists have weird tastes.

So the iron, the lava, and the pot pie all glow the exact same color of red at 1000°C. Why? No one knew why before 1900. In fact, those poor saps *couldn't* have known. That's because the phenomenon can't be explained with pre-1900 physics, which is now called *classical* physics in analogy with *classical* music, which also couldn't explain what the hell was going on. But while everyone else was partying like it was 1899, Max Planck was mounting a revolution. (Max *Plonk*. Pronounce it like you are mocking fancy English and not like you are a Midwestern NASCAR snob.)

It was known for some time that light was a wave of electromagnetic radiation. I know, that's a mouthful of science jargon. But there are really no other words for it, which is surprising because it is not difficult to picture. Imagine taking a ball with an electric charge and shaking it up and down. Take a balloon and rub it in your hair, for example. In addition to looking a bit foolish, you are creating a wave of electromagnetic radiation. You are basically like a small radio tower, which shakes electric charges up and down a few thousand times per second. How fast you shake your charged ball is called *frequency*. Remember that. Frequency. Say it out loud because it's going to be important. Did you do it? Come on, do it. Did you? Waiting...did you do it yet? Okay, fine, I'll stop.

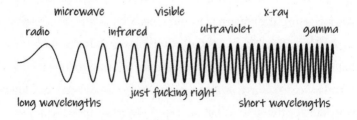

6

The frequency of light determines its color. The frequency can be any number, but we generally divide them up into a few categories to give them names. People like naming things, and scientists are no different. Really small frequencies are radio waves. As we increase the frequency come microwaves, then infrared, then red, orange, yellow, green, blue, indigo, violet. Next is ultraviolet (UV), then X-rays, and finally gamma rays, which have the largest frequencies we have given a name to. All those are "colors" of light that are vibrating electromagnetic waves traveling around at a million kilometers per hour. A tiny fraction of it can be detected by your eyes, but the vast majority of it is invisible to us. If we could see it, it would be quite a trip, like living in a pre-1960s Disney movie during the inevitable scene where someone gets drunk or high. Nature is fucking lit, isn't it, Dumbo?

Everything with a temperature glows with a certain set of colors. Even you glow, albeit in infrared light. That's right, your partner next to you in bed is a source of radiation. Better sleep with a tinfoil hat on tonight. So Planck's task was a simple one: find an explanation for how something hot could create these very specific colors or, in the parlance of waves, frequencies. Easy, right?

With immense ingenuity, he found the solution. Actually, he made a wild guess and it kind of worked, and he didn't like it but told other people about it anyway, and they didn't like it either. But with that hack, he birthed quantum physics. Next time you hear the phrase "got it down to a science," remember that most of what we do is hack at ideas until they work. Science. It...kind of works...bitches.

You may wonder, *if it worked, why didn't Planck and his contemporaries like it?* Well, we have to remember that 1900 was a different time. People then generally liked the idea of a single source of truth—rather than today when social media lets everyone invent their own reality. Planck didn't want to be a radical who hacked together ideas. He and everyone else wanted an explanation that fit within the existing understanding. But can you really blame them for wanting to feel comfortable?

Quantum Leaps... Like the TV Show but with Real Science

Hack. That's a funny word. Hack. Say it out loud. I promise the people around you will start laughing. So what was this hack of Planck anyway, and why call it a hack? In a nutshell, Max Planck said the oscillations creating the light from the hot object could only be at specific frequencies instead of any frequency. It's a *hack* because it is completely arbitrary. It's like saying you can only eat your ice cream in spoonfuls instead of licking any amount you want off it. (I'm with Planck here, by the way. Eat your fucking ice cream. We don't have all day to watch you lick it to death.)

Constantly thinking about rapidly oscillating things is annoying, so let's talk about energy instead. There are basically two forms of energy: kinetic and potential. Kinetic just means moving. Moving things have energy. It's the essence of the concept. But nonmoving things can also have energy if they have the *potential* to move—hence the name. This is all well demonstrated by an example. I've picked out a good one: an oscillator! Remember when I told you about physicists and their patterns? Well, their favorite pattern is a steady oscillator. Why? Mostly laziness. A steady oscillating motion can be turned into mathematical equations that can be easily solved. You cannot imagine how important this is to us. Most of our equations cannot be solved at all, so to find an example that can be solved is a gift.

Oscillators are everywhere: kids on swings, pendulums on clocks, pistons in a car engine, wings on a hummingbird, strings on a guitar, my general attitude toward most people before and after watching the news, and so on and so forth. Which one are you thinking about? The hummingbird. I knew it. You are imagining those cute little wings flapping like mad. But hummingbirds move their wings too fast for us to get a good sense of the fact that, at least for an instant, the wings aren't moving at all. What?! Lies. How can that be?

Think about a child on a swing instead. When a child on a swing approaches the top of their swing, they are slowing down and, in fact, come to a complete stop. Then they yell, *push me higher!* And you're like *nothing would please me more than the monotony of this pointless task.* But—bonus!—while you're pushing your kid (or some other kid if you don't have one), you can think about physics. At the top of their swing, when they are not moving, does the child have energy? Yes! This is *potential* energy. It's energy in wait. It's the energy of a loaded spring or battery. It has the *potential* to create motion even if it is not moving. A bowl of cereal on the floor has no potential energy, but a bowl of cereal on the edge of the counter annoys the fuck out of me. I mean, really, you have the whole counter, and you chose to put the bowl there—do you know nothing about potential energy?

While the child is swinging back and forth—and you are going out of your mi...err...thinking about physics—their energy is changing back and forth from kinetic to potential. But here is the critical part: the total energy—the kinetic plus the potential—is constant. You may remember the mantra of elementary school science class: *energy is never created nor destroyed*. This, by the way, is the essence of a good scientific idea. It takes something complicated and makes it simple.

Every steady oscillating thing has energy, and while that energy may change its *type*, the total energy stays the same. So instead of thinking about Planck's idea in terms of frequency (color) of light, we can think of it in terms of *energy*, a much more familiar concept.

What does energy do? It *flows*, right? A child's energy on a swing flows from kinetic to potential, just like their motion is a smooth process. When things move smoothly, we call it *continuous*. We expect energy to be continuous. Wrong!

Planck discovered that energy is not *continuous* (smooth, like the correct type of peanut butter) but *discrete* (chunky, like "all-natural, chemical-free" hipster peanut butter—damn you millennials!). While everyone else was trying solutions to the problem, allowing any ol' value for the energy, Planck said that the energy could only come in chunks. The tiniest fragment of energy is a quantum. Hence the name. Remember that for the next trivia night.

Planck maintained that this idea was only a mathematical trick and not to be taken seriously. However, one badass physicist not only took it seriously but also took it to the next level. You might have heard of him. His name was Albert Einstein.

Einstein was also thinking about light and used Planck's idea. The result was his revelation that light itself came in chunks of energy, which we now call *photons*.

Nowadays, we know that each of the four fundamental forces is carried by one of these quantum chunks of energy we sometimes call *particles*. That's it. That's quantum energy. I'm going to put it again in a box, so you remember.

> In classical physics, energy is continuous (smooth). In quantum physics, energy is discrete (chunky).

That Doesn't Sound Very Impressive

I see you are not impressed. Let me help you out. Before Planck, there were only two fundamental constants of nature: G and c. I'm not going to tell you what these are. Just kidding. Of course I am. Wow, you are really falling for it. G is the *gravitational constant*. Strain your brain for a moment and try to think back before iPhones and Facebook, all the way to high school. Remember when your science teacher told you about Isaac Newton and the famous apple story? What did he discover again? Not to stand under trees. No, he "discovered" gravity, as if it wasn't obvious.

Of course, Newton didn't "discover" gravity but rather a framework for understanding it systematically. What you probably completely forgot—if you were even listening—was the formula Newton came up with:

$$F = G\frac{m_1 m_2}{r^2}$$

Math. Yes. Fucking right. I love it. I know, I know, you hated math in school or whatever. But I have to listen to you drone on about your yappy little dog and keto diet, so you are going to listen to me about this one tiny math equation for five goddamn seconds.

The thing on the left, F, is the force of gravity—it's what you want to calculate (if you are into rocket science or just want to throw a rock at science). Those m's on the right are the masses of two things—*any two things*—like the Earth and Moon, for example, or a rock and the Earth. The r at the bottom is the distance between the two things. You have to square it before you divide. (BEDMAS! Ah, see, it was helpful.)[4] The masses and distance are known. But this was Newton's genius: he said that no matter what the objects were or how far apart they became, one single number, G, would allow you to calculate the force. Now that's fucking cool.

The other fundamental constant of nature is c. It's the speed of light. It's, like, fast and shit.

What Planck discovered was more akin to Newton's constant, and it is aptly named *Planck's constant*. It is given the symbol h. Planck and his pal Einstein even had a similar equation:

$$E = hf$$

Ohhhh, look at it. It's even more beautiful than Newton's! It says that energy is directly related to frequency. To calculate it,

---

4 In case you forgot: brackets, exponents, division, multiplication, addition, subtraction. BEDMAS. It's the order in which you ignored your homework.

you only need a single number: Planck's constant. Dope! In case you were wondering, and I know you were, Planck's constant has been measured to be really fucking small. More technically, it is 0.000000000000000000000000000000006626. So yeah, really fucking small. This checks out, though, since we wouldn't expect the smallest unit of energy to be large. Otherwise, we probably would have noticed it sooner.

In classical physics, the energy of a steady oscillator can be any amount of joules. But in quantum physics, the energy must come in units of hf. If some steady oscillator has a frequency proportional to h, then the energy of that oscillator can only be $1h$, $2h$, $3h$, and so on. These are the quanta, the chunks of quantum energy. A common way scientists talk about this stuff is using the idea of a ladder. The amount of energy is the height of the rung. But energy can *only* be on the rungs—it's discrete.

This is the way Niels Bohr first conceived the quantum nature of atoms. Atoms were once thought to be the smallest units of stuff, but by firing light at them, we found out that even atoms have internal guts: protons, neutrons, and electrons. Protons and neutrons live in the center all bunched up, and electrons fly around them in circles. It looks like a solar system, and it is indeed called the *planetary model*. Bohr's idea was that the electron could only be on a discrete set of orbits, just like the ladder picture. The only difference is that we call the allowed amounts of energy *levels* instead of rungs.

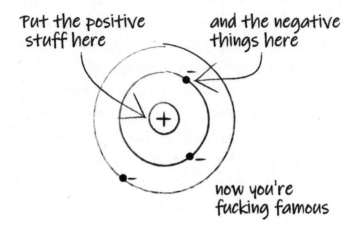

Electrons in atoms are not quite steady oscillators, so the picture becomes inaccurate. In fact, it is far more complicated depending on how deeply you want to understand it or how accurately you need to predict its behavior. But the essential quantum feature is always the same—*energy is discrete in quantum physics.*

How does an electron change energy levels? It absorbs and emits energy quanta in the form of photons. And the picture is complete. Energy is never created nor destroyed in the quantum process but moved from atom to atom as quantum chunks of light.

Cool, Cool. No Doubt, No Doubt.
But How Do I *Use* Quantum Energy?

Do you have pain? Does your water taste funny? Do you want more money? Are your chakras, like, not what they are supposed to be like or whatever? Then quantum energy can help you...for the low, low price of four easy payments of $49.95. Skeptical, right? You should be because... BUT WAIT! There's more. If you act now, you'll get two quantum energy thingies for the price of one. Well, now we're talking!

I would really like to meet someone who falls for these scams. Actually, scratch that. I don't think I would. But if you are one of them, I have good news, and I have bad news. Which would you like to hear first? Trick question. That's not how books work. Get your head in the game. The bad news is that I don't have a quantum energy thingy that can make you healthy and rich, but you can still send me $200 if you really want to. The good news is that you are reading this book, and when you finish it, you will no longer fall for these scams, and then you too can get outraged when you read quantum bullshit on the internet. What a life you will soon have.

Quantum energy bullshit comes in many forms. In most cases, however, the "quantum" part seems to be entirely superfluous. I know, I sound surprised, don't I? But if you were going to go through the trouble of labeling your product as using or producing "quantum energy," the least you could do is emphasize its discreteness. Every quantum energy woo peddler I've come across is using the classical notion of energy with the word *quantum* slapped over the top of it. Where the hell is Planck's constant? If quantum healing crystals were really quantum, they'd be quantum dust particles. Now, there's an idea...

In the vast majority of cases, the scam disappears as quickly as it arrives. Some sounds-too-good-to-be-true YouTube video sends you to a bogus website, which sends you through various other poorly designed portals until eventually one of them asks for your credit card info and you've forgotten exactly what it was you were buying into, but somehow you ended up with hepatitis. Like meeting a used car salesman with gold teeth, red flags should be flying up all over the place. Don't be a fool. Delete all your accounts, and start a fresh browsing history with Bob Ross videos. You should be safe for about an hour or two.

As far as actual products go, the most common are the ones supposedly *imbued* with quantum energy. Some products are entirely harmless to use—like, say, a "quantum" crystal that sends you positive quantum vibes. No harm there. We all like positive vibes. It's like eating sand—a bit salty but not going to kill you. However, if you eat sand as a replacement for food, then you are in trouble. And if you use "quantum energy" products as a replacement for conventional medicine, food, or other basic necessities, well, let's just say the *continuity* of your life might end up *discrete* a lot faster than you expected.

Take, for example, the Quantum Xrroid Consciousness Interface (oh, it's real)—a device now banned in the United States, created by a fugitive wanted for fraud who still sells the fucking thing from an old hotel in Budapest. It's hard to say what this or any other alternative medicine is claimed to do since the pitch always changes. Headaches, stress, high blood pressure, viral infections...cancer? Sure, why not? Cancer. Whatever complaint you might have, quantum healing bullshitters are not below telling you they have the solution.

While many of these devices have been produced, they are all different and make different health claims. Often, the same product will resurface in several places under different names. Case in point: the Quantum Xrroid Consciousness Interface is now called the Electro Physiological Feedback Xrroid, or probably something different by the time you read this. I'm just going to call it the Xrroid thing, since that seems to be a commonality and doesn't lose the over-the-top ridiculousness of it.

There are some things all electronic "energy medicine" devices have in common though. They all have some electrodes that need to be connected to the body—usually the head because, you know, that's where the mind or consciousness is or whatever. There will be some monitor that reads electrical signals and interprets them in bullshit language like "quantum energy profiles" or something similar. Then—the really expensive ones anyway—send electric pulses back into your body. Don't worry though; they are as harmless as they are useless. To be clear, they are at best placebo. As a bullshitter, you only want your victi...err...customers to *believe* something is happening. Actually having something happen could be dangerous, and you'd be outed before your scam even got going. Of course, in the context of medicine, doing nothing can actually be harmful.

Apparently, you can call someone who creates a scam product built from RadioShack parts an "inventor." So the Xrroid thing "inventor" was Bill Nelson, a self-styled "professor" who later gave presentations of the device as Desiré Dubounet, an alter ego. It seems to be remarkably common for scammers to change the name of not only their product but also themselves routinely. Anyway, at one point, they were recorded stating that after only a few "treatments," a patient's tumor literally fell out of their body. I mean, do you laugh, cry, cringe, or what? These stories always reach the point of insanity such that you'd never guess, over the long history of the scam, there are over twenty thousand Xrroid things out there, including in a U.S. hospital for fuck's sake! In some sense, it is fascinating enough that I won't be surprised when the documentary pops up on one of the seventy-eight streaming services I've signed up for.

I think just reading about the existence of this shit is giving me a tumor. But rather than put a vibrating toy hat on my head for a few hours, I think I'll see a real doctor. By the way, in the case of the Xrroid thing, there have been documented cases of people dying from using it instead of seeking advice from a medical professional. Dying is generally not considered a positive health outcome, which is probably why the device and its creator are banned from the United States.

Enough! Tell Me the Real Secret of Quantum Energy!

As a good rule of thumb, it's probably best not to use *any* product claiming to have or use quantum energy. Either it is complete bullshit being sold by a fraudster, or it is something that produces measurable quantum energy using radioactive elements. Yeah, *radioactive*—you know, the thing that is often indicated in the universal language of giant warning signs. That's *real* quantum energy.

Remember the discrete packets of energy we talked about, the quanta? Well, just because they are discrete and sound "small" doesn't mean they can't do serious damage. Ever wonder why you wear a lead plate on your chest when getting your teeth X-rayed at the dentist? Ever wonder why they have extra lights that shine right into your eyes? Ever wonder why they pick at your gums until you bleed? They are just mean, that's why. But at least one of those things is necessary, and that's because X-rays have a lot of energy.

Each X-ray is a quantum (a photon) of the same amount of energy. More or less of them don't change the *kind* of damage they can do to your body. Long exposure can do more total damage, but that is just a buildup of each photon doing its atom-bashing thing. Over one hundred years ago, Marie Skłodowska Curie discovered radium and introduced us to the wonders and dangers of radioactivity. Radioactive elements just naturally emit all sorts of high-energy quanta, including photons. But they don't do so in a controlled way like an X-ray machine at the dentist's office or airport. Exposure can be disastrous and deadly. Of course, we didn't know this a hundred years ago, and we put the lovely glowing stuff on everything. This *real* quantum energy was even used to make energy drinks—literally radiation water. Maybe you read that too quickly, so I'll say it again—companies used to sell water with radioactive material dissolved in it. You don't want to know what happened to people addicted to that stuff. Their faces fell off. Why did you read that? I told you, you didn't want to know.

The moral of the story is you don't want real quantum energy. Be happy that consumer protections exist to shield you from quantum energy and direct you toward totally safe and definitely not addictive drugs like caffeine. I, for example, could stop drinking coffee whenever I want. I just don't want to right now, okay?

All this talk of people's faces falling off is depressing. Tell me something positive about quantum energy.

Cancer

Oops. What I meant was *killing* cancer. What kills cancer? That's right. Quantum fucking energy. You may have heard of radiation therapy. It works in basically the same way as radiation water by killing living cells in your body. However, when you put your life in the hands of medical practitioners and scientists, they can make sure the radiation only kills the cancerous cells. This is a precision science that has taken many years and many experts to perfect. The idea that some rogue entrepreneur has created some magic solution that "doctors don't want you to know about" is the stupidest thing I've read on the internet, and I use Twitter.

Quantum energy has given us more positive things as well, like the term *quantum leap*. We already know how electrons move between energy levels—they absorb and emit photons. Without reference to the photons, we often say the electrons "jump" between energy levels. Nowadays, the term is more often understood idiomatically to mean a discrete change in anything normally expected to change continuously. There are "quantum leaps" in economies and technology in addition to the energy of electrons. The fact that this is the only figure of speech whose literal meaning refers to a scientific concept is both amazing and depressing. It's amazing because it shows the popular influence of quantum physics but depressing that we don't have more science-inspired idioms. We should definitely have more. How about we start using *photosynthetic* to figuratively mean that warm feeling on your skin when you are in the sun. No? You don't like it? I don't care. I'm using it.

Quantum energy is the fuel that powers all microscopic processes. It's always there. So if you want to build things at the microscopic scale—like tiny electronic chips for example— you definitely need to worry about quantum energy. Most of us aren't microelectronic engineers though. In fact, I've never actually met one, but they do live among us! For the rest of us, individual particles are nothing to be concerned about unless you really want to understand deeply the processes that govern the universe.

Yeah, I knew I could suck you in with that line. Here we go. I will tell you one cool thing you can use quantum energy to explain in everyday life: the sky. More specifically, its color. Spoiler: it's blue, except when it's not. But it is usually blue when you look at it. So why is that?

Common answers to *why the sky is blue* include: the ocean is blue and reflects blue light; water is blue and there is water in the sky that we can see; and blue light reflects off the dust and stuff in the sky. These are all incorrect. But we have all the tools needed to understand it, and it comes down to quantum fucking energy!

Remember back in the 1800s when the old dudes didn't know why shit glows when it is hot? It was a few pages back if you forgot. Anyway, at 1000°C, stuff glows red. When it is 3000°C, it glows yellow, and approaching 6000°C, it glows white. Beyond that temperature, things start to look blue. Is the temperature of the Sun so high that it glows blue and lights up the sky with its blue light? Nope. So you might guess then that the temperature of the Sun is around 3000°C since it looks yellow. Nope. Go outside at midday and take a photo of the Sun. DO NOT LOOK AT THE SUN. Someone is going to do it anyway. I just know it. For those that didn't blind themselves, what does your picture look like? White. The Sun is white. That's because it is 5500°C.

Now, if you remember your color theory lessons in art class, you'll know that white is not a color. In fact, most things we call color are not colors in the sense that they are a frequency of an electromagnetic wave. Pink, for example, is not a single frequency but many frequencies. So when you see pink, you see many photons with different frequencies, and the meat sack that is your body can't tell the difference. This is not the place for a lesson on human color perception, but know that we are pretty pathetic sensors of nearly everything. When the robots take over, they won't have pink because they will use sensors that detect all the frequencies present in light. Speaking of all the frequencies, that is what white light is. White light is the combination of all frequencies.

The Sun is basically a big ball of hot gas spitting out photons of every frequency in all directions. The energy of each photon is given by our trusty equation $E = hf$. If that energy matches an allowed energy level difference in an air molecule, then the photon is absorbed. When it is emitted, it does so in a new random direction. This is aptly called *scattering* and is why you see photons that aren't in a direct line from the Sun. As it turns out, the energy level difference in air molecules is better tuned to high frequencies like blue, and quantum physics does its discrete energy trading thing.

So blue light is being subtracted away more from the direct sunlight than greens and reds. When it is safe to look at the Sun, during sunrise and sunset, the photons must pass through a lot of atmosphere. That provides many opportunities for a blue photon to find a molecule and scatter through the absorption and reemission of energy. What's left that we can see are the reds and greens, which together appear yellow. If you look up, though, you'll find those quanta of blue energy scattered back toward you.

Get Rich

That's swell. But can quantum energy actually help me get rich or what? If you are going to ask stupid questions, I will give you stupid pedantic answers. Yes, quantum energy can help you. In fact, it already is. Atoms, photons, and all the other quantum stuff are not different from everyday stuff like tables, iPhones, and Pierce Brosnan. (He's still relevant, right?) All matter—you, me, Pierce, hipster beards, coffee, even bullshit energy crystals—is made of atoms. And all light is made of photons. Everything that happens is ultimately due to quantum energy at some point. All is quantum. Embrace it. Bask in it. And then say *fuck it, who cares.*

Yes, electrons are jumping up and down, absorbing and emitting photons right in front of your eyes. And this is happening all around you on a massive scale. You just can't see it. Ironically, though, this actually makes it far less relevant to worry about. There are a billion billion billion atoms in your hand. So thinking about the quantum nature of your hand won't help you when you decide you need to move it toward that cup of coffee. Mmmmmmm coffee. Totally not addicted to it.

Quantum physics is interesting because it pushes the boundary of what we experience in our everyday life. We owe so much of our understanding of nature and technology to this simple idea that energy is discrete, contrary to what our senses would tell us. But even for a quantum physicist, the vast majority of an individual human's problems are defined on human scales—not microscopic scales. This is obvious when you realize that most of your issues are literally other humans.

So suppose then that you find yourself in a situation where

someone is telling you something about quantum energy. In that case, they are either an expert—and things are about to get scientific and technical—or they are full of shit, in which case you should run. Maybe you should run in either case.

2

Fucking matter waves

Picture a tropical paradise. It's sunset. The waves are gently washing up on the shore. You feel the sand between your toes. It's also trapped in your hair, and how the fuck did so much get in your swimsuit? Deep breaths. Listen to the waves. Match your breathing with the waves. In. Out. In. Out. In...ou...aww, what is that fucking smell? Dead fish? This beach is the worst. Whose idea was it to come here anyway?

There's something about waves that primes people for believing pseudo profound bullshit. Don't believe me? Look at this:

Success ripples out from the reflection of your own essence.

I just made that up. You can't even google it. It doesn't mean anything. It's complete bullshit. But it feels *so* profound, doesn't it—especially with that rippling water? Mmm, yeah.

But there *is* something special about waves. No, I'm not going to give you some scientific license to get overly emotional with your cosmic connection to the sunset. I mean, I'm not a robot—I take pictures of sunsets too. I just don't allow the emotional response to seeing the sunset or hearing the waves cloud my rational judgment. But if the ocean *does* enchant you, get ready for the real kicker. Waves are not limited to crowded tourist traps. *There are waves everywhere.* Everywhere. Even inside you. Yeah, total mindfuck.

In the last chapter, we learned about energy and that it gets moved from one place to another, powering all aspects of life and the inner workings of the universe. But *how* does it do that? The answer is waves. Waves are things that carry energy from one place to another. So what if you could tune in to the frequency of these waves, effectively syncing your energy with the universe? Ah! Gotcha! Now that would be magical fucking bullshit.

There are a lot of things you can do with waves—surf them, listen to them, send dic...err...wholesome cat pictures with them. You can also pretend to perform supernatural feats with them. But then you would be lying. However, if you are going to lie, you might find the following table of jargon useful, so at least you sound smart.

Property	The fuck is it?	Measured in
Wavelength	The distance over which the wave repeats. For example, the distance between two crests of ocean waves might be 10 meters.	Meters (m)
Frequency	How many waves pass in one second. For example, if one water wave hits you every second, the frequency is one per second.	Hertz (Hz)
Wave speed	How quickly one crest travels. Small ocean waves travel at about 10 miles per hour or 4.5 meters per second.	Meters/second (m/s)
Amplitude	How high the crests are. Typical day ocean waves on a surf beach are about 1 meter high.	Meters (m)

No jokes here. This is a legit table of fucking facts.

A Completely Fictitious Prehistory of Waves

If you asked a random person what a wave is, you'd always get an answer, but that answer will depend on when and where you asked it. Ask it in Western Europe, and you'll get a nonverbal hand gesture signaling a greeting. Wave the same way to someone in Southern Europe, and you'll be giving an insult. Ask it two hundred years ago, and you'll get an army salute. Ask it two thousand years ago, and you'll get back two legs of lamb and a loaf of bread sacrificially waved at you—part of which might have to be burned in a fire or something. Ask a sailor, and they will describe the motion of the surface of the ocean. Ask a New Yorker, and they'll tell you to fuck off.

And that's pretty much the entire nonscientific history of waving. The point is, a *wave* is a term with many meanings, but there is a common thread in all of them: *back and forth motion*. The motion of what, you ask. Anything. Anything? Any. Thing. And sometimes nothing.

Music to My Ears

As with most things, we attribute the origins of the systematic study of wave motion to the Greeks. This just goes to show how important it is to take good notes and write things down. Just imagine in another few thousand years when someone finds this book and is in awe of how visionary twenty-first-century humans were. Well, me anyway.

But what were the Greeks doing twenty-five hundred years ago? According to Hollywood, Greeks were into wearing togas, politics, fighting bare-chested with awesome abs, and music. If you are out late at night in Athens, you might still see a couple of these today.

Let's do an experiment. Pick up the lyre next to you. What's that? Don't have a lyre? Sigh. A guitar will do then. Still no? Ah well. Go find a store selling musical instruments, and pretend to test one out. Say it's for a friend—no one is going to believe you have the commitment. Pluck one of the strings. You know what, fuck it. At this point, I'm under no illusion that you are actually going to do it, so I'm going to get pretty descriptive here. What is going to happen? First, sound. That'll be the most obvious effect. Then you will notice something about the string. It vibrates.

The ancient Greeks may not have had toilets, but they definitely didn't stink at noticing coincidences. They realized that it was the vibrating of the strings on musical instruments that created sound. Pythagoras went further and described the relationship between pitch and the length of the string on his sick lyre. Look at your guitar. It has a hollow body to amplify the sound created by the strings, which extend up the neck to the head. Along the neck are the metal bars called frets. Pushing the string down against a fret does nothing more than shorten the string. Shorter string, more badass guitar solo—and a higher pitch. Once you know that the length of the string determines the pitch of the sound, you can play any John Mayer song. Tuning the guitar is not necessary.

But if you did want to tune a guitar, how would you do it? Probably start with YouTube. I used an app. Now hold up. Before I give a false impression that I am a guitarist, I want to admit here once and for all that I could only ever play one song, and I only practiced it long enough to convince a love interest that I had more than one talent.

Pythagoras didn't have YouTube, an app, or clean underwear. But he did have one thing that can replace all of those: mathematics. He was pretty famous for it—not the dirty underwear thing, the mathematics. I won't bore you with instructions on tuning a guitar with millennia-old mathematics (again, YouTube is good for some things). The key is this: pluck a string, and it will vibrate. It will vibrate more or less the same way no matter which way you pluck it. That's the *natural frequency* of the string. In music, it's called the fundamental tone, or first harmonic, of the string. It vibrates at some frequency, say middle C, or 261.6 Hz. Pinch the string to half its length, and it starts to vibrate *twice* as fast, now tenor C, which is...wait a sec... The trusty Casio calculator says 523.2 Hz.

What does the number 523.2 Hz mean? Remember Hz is hertz, cycles per second. The guitar string for middle C is moving back and forth 261.6 times every second. I know! Fast as shit! It does this every time you pluck it, never wavering, always 261.6 Hz. That seems oddly specific. What the fuck is up with that? There are two possible answers—you guess which one is bullshit.

1. The physical properties of the string—its length, tension, and mass—determine, through the fundamental laws of physics, how it will vibrate.
2. The string has built into its very being vibrational tendencies, which can be ignited by cosmic energy in a harmonic symphony.

It's just too easy to make this shit up.

The mathematical structure of how many different tones come together to make music is fucking beautiful. Understanding the science behind it might seem like a dry academic exercise, but it doesn't take the beauty out of it. In fact, it's quite the opposite. When I see a guitar string vibrating, I see waves, a source of energy that is transferred by another wave to my ear, which contains a tiny drum that starts to vibrate, causing electrical signals to be sent to my brain. From there, all sorts of chemical and electrical interactions create more waves, which cause emotions and spark memories...maybe even childhood memories, memories perhaps of my grandfather playing guitar for his grandchildren. Kidding! He hated children. I only remember once getting a check from him on my birthday. I think it was to buy food for the cat I stole from his apartment—long story. Never mind.

Thanks for Tuning In

Plucking a guitar string leads to the obvious reaction of it vibrating. Push something, and it moves, duh. But why does it make nearby strings move as well? And why does it make my eardrum move at the same frequency? The answer is *resonance*, and it is incredibly important to understand how it works if you want to shield yourself from enormous waves of bullshit.

Recall that the vibrating of the string happens at the natural frequency. There is nothing deep or meaningful about that frequency. It's decided on by the physical properties of the string—its length, tension, and mass. Change the length, change the natural frequency.

Now check this shit out. If you yell at your guitar—that's right, yell at it—the strings start vibrating. What? Seriously, they do—I tried it. It's so amazing that people around you will pull out their phones to record it. Though now that I think about it, they may not have been impressed by the strings. *Physics professor yells at string. You won't believe what happens next.*

There's nothing special about a guitar here. Yell at anything that vibrates—guitars, drums, wineglasses—and if your voice is as sweet as mine, you'll get that thing to vibrate. The key is to yell at the natural frequency of the thing. That's *resonance*. In other words, if you yell at some frequency, everything around you that is capable of vibrating at that frequency will start to do so. They resonate.

Resonate. This word you are surely familiar with since it's been stolen from physicists to now figuratively mean *be in agreement with a thought or action*, which—as far as analogies go—is not bad. However, in the age of information, you can always find someone angry about anything on the internet, and the overuse of the word *resonate* is no different, apparently. Ben Zimmer (*New York Times*) writes, "No matter what your line of work is, it's best to use *resonate* sparingly if you want your words to fall on receptive ears."[i] Ugh, language purists... At least I have good reason to be grumpy. So fuck that, we are going to resonate the shit out of this chapter.

All this yelling at guitar stuff is exactly the opposite of what you do when you tune a radio, or what you *did* to tune a radio twenty-five years ago. Today, of course, a computer does it for you. What's happening with a radio broadcast is that someone else is yelling, literally in the case of conservative talk radio. That yelling is turned into radio waves and sent out in all directions. In fact, you are being bombarded with radio waves right now. But you aren't *tuned in* to them. By changing the dial on your old radio to match the frequency of the broadcast, you create an electric circuit inside the radio with the same natural frequency—your radio *resonates* with the broadcast. This is what *tuning* a radio means.

Quantum?

Umm...excuse me. I thought this book was about quantum physics? Calm down. I'm getting there. In fact, in making the connection between sound waves and radio waves, we're just where we left off. Radio, if you recall, is electromagnetic radiation—light. It has a frequency, which is defined to be anywhere between 30 Hz and 300 GHz (300 *giga*hertz or 300 billion Hz). My favorite radio station growing up was CIMX-FM (88.7 FM), which was a progressive rock station out of Windsor, Ontario, branded as 89X. Those weren't random numbers, and the name didn't come from the fact that you had to be the eighty-ninth caller to win the tickets to the Depeche Mode concert. The number 88.7 referred to the frequency of the broadcast, namely 88.7 MHz—that is, 88,700,000 hertz. Tune your radio to that natural frequency, and you'll be singing along with the Cure in no time.

I don't care if Monday's blue, Tuesday's great, and Wednesday's shit... I just realized something. I didn't tell you *why* a radio wave is even a *wave* at all. We're going to take a quick detour back two hundred years or so. But I promise we're going get to that sweet, sweet quantum goodness soon.

Detour: The Physics Experiment to Rule Them All

Recall that radio is just one type of electromagnetic wave among infrared, X-rays, and even visible light. To a physicist, all electromagnetic waves are light, even if it can't be seen with our eyes. Hell, many of us spend most of our time in dark rooms with equipment that "sees" all the light we can't see with our eyes. Of course, before we knew all the details about invisible light, we had quite a bit figured out about the colors we can see, including the fact that it vibrates—in other words, it is a wave.

The year was 1801, one hundred years prior to Planck and the quantum hypothesis. People didn't know if light was a wave or a particle. But one experiment was so convincing that it instantly solidified in the minds of most scholars that the nineteenth century was miserable as fuck and they should start using science to make life easy and comfortable for the most ungrateful of people—generally, most people. Well, that *and* that light was indeed a wave! The experiment has been dubbed the "double-slit experiment," and while esoteric sounding, it is quite descriptive.

Now, you are going to be a bit underwhelmed, but remember: this was 1801. You would have been too busy not starving to dream up the multibillion-dollar, thirty-kilometer-wide particle accelerators we have today. Instead, what you would have done is close the shutters, poke a hole in them to let a beam of light through, and direct it toward a card with more holes in it. If done correctly, congratulations, you're famous because you just proved light *interferes*.

Here is what the "apparatus" looks like:

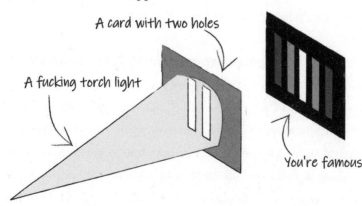

A card with two holes

A fucking torch light

You're famous

Light hits the card with two slits cut out—hence the name "double-slit experiment"—and ends up illuminating another card behind it. The important part is that light on the final card is a wave pattern. It is brightest at the center, between where the slits were, which could only mean that the light "adds up" there. When the peaks and troughs of a wave (light, water, sound, whatever) add up or cancel, it is called *interference*. This behavior is the hallmark of waves. In fact, by playing with the size and distance of the slits, you can find out that the pattern behaves exactly as waves would. Of course, it's not definitive proof that light is a wave, but you'd be hard-pressed to find an experiment more convincing than this.

Given the simplicity, you might be wondering why I'm telling you this. I mean, you probably just believed me that light was a wave anyway. Well, first of all, you should be skeptical—at least a little bit—about everything you are told. Who is telling you this? What do they stand to gain or lose by you believing or not believing it? Sometimes the stakes aren't very high. Other times, you have to call loud and clear: *bullshit!* I don't really have anything to sell you since you already bought the book, so I have little reason to lie to you about the science of light. The real reason I am introducing you to the double-slit experiment is that it is still used today—not in the bedrooms of affluent nineteenth-century Englishmen but in quantum physics laboratories, the bedrooms of twenty-first-century physics graduate students. In fact, no matter where you are or what time you are reading this, it is midnight somewhere else in the world, and there is a poor grad student in a lab performing a modern version of the double-slit experiment right now.

In that lab, the student performs some variant of the quintessential demonstration of what has been called the most irritating feature of quantum physics. I give you...

Wave-Particle Duality

Earlier I said that light has a frequency—its color. But I also said it comes in little packets of energy—photons. Actually, Einstein said it. But he's dead, so I'll take the credit. Anyway, photons—those are particles. Particles, as we like to imagine them, are things with a definite location. You're probably thinking of a little ball or a grain of sand. That's good. It has a position in space we can all point to. Imagine us all in a circle pointing to a ball. Boy, would we look stupid. Not because that would be a weird thing to do but because light is a *wave*, as the double-slit experiment clearly showed us only a few moments ago.

A wave, like an ocean wave, doesn't have a location. If I ask you where the wave is, you'll rightfully be confused. You can *wave* in its general direction, but you can't point to where it is. So waves and particles seem quite the opposite of each other. But now we have a problem, a big fucking problem. Wake up! Because we agreed on one hand that light is a wave and, on the other, that light is made of photons. A contradiction or—if you want to sound profound—a *paradox*! OoooOooooO!

Maybe, though, light sometimes behaves as a wave and other times as a particle. Perhaps, depending on the situation, it is always one or the other, and physics will tell us which. Surely, *surely*, it can't be *both at the same time*, right? Wrong! It's time for the double slit to make a comeback.

Imagining performing the double-slit experiment again. But

this time, imagine the light source was so dim that only a single photon went through the slits at a time. Suppose one photon arrived every second or every hour or every year—it doesn't matter. From experience and intuition, you would expect that the light intensity would build on the card immediately behind the slits because particles, being like balls, just travel in straight lines.

But this is not what happens. Instead, one by one, the photons hit the card in a seemingly random way. After a while, though, you notice something. The photons are making a pattern—exactly the same wave pattern you'd expect if light passed through both slits simultaneously. So in a single experiment, light behaves as *both* a particle and a wave.

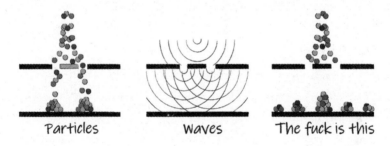

Particles   Waves   The fuck is this

If you have followed along this far, you'd agree that this is weird. Why can't it be one or the other? Stupid nature has to go and ruin everything. This paradox really *really* bothered the folks who developed quantum theory a hundred years ago. But just because a bunch of war-weary, midcentury Europeans were confused doesn't mean you should be too.

Now, here is where I diverge from many of my fellow science communicators. You'll see in countless popular science articles, YouTube videos, documentaries, news stories, and even academic journals the statement *light is both a particle and a wave at the same time.* We even invented a name for it: *wave-particle duality.*

I don't like it, not one bit. But physicists are trained to solve equations and shit, not to be linguists or poets. We're not good with the words. So here is how I want you to think about such paradoxes. First, let's rephrase the apparent contradiction one more time. In the double-slit experiment, light has both wave-*like* and particle-*like* behavior. It can't be both a wave and a particle—that's a contradiction. So what fucking gives? Let's not toss our hands in the air and say *it is both at the same time, magic and mystery!* Let's instead say light is *neither* a wave *nor* a particle—it's something else, something new. Let's just call it light and be done with it. Cool? Cool. Comfortable?

Not So Fast

So light is not a wave, but it does have *wave-like* qualities—each photon has a frequency, for example. Remember the Einstein-Planck equation. Well, we can do a little rearranging to see

$$ E = hf \implies f = \frac{E}{h} $$

that the frequency of light is the energy divided by Planck's constant. And this doesn't really bother us because we have seen the double-slit experiment. Fine.

But then, in the 1920s, Louis de Broglie came along with the idea that not just light was wave-like but matter as well. Now, when you read about de Broglie, it is always emphasized that he was a physicist *and* an *aristocrat.* Ooo la de da! Actually, I'm not sure why the latter part is relevant, but my only understanding of aristocracy is from the 1970s Disney film *The Aristocats*. Good film. Anyway, Louis the aristocrat fancied himself a deep thinker, and what better way to prove it than develop a radical new conception of all of physics and reality? Probably, if it weren't for the approval of Einstein and Schrödinger—who were proper science celebrities by then—de Broglie's ideas would have remained obscure. Instead, he won a Nobel Prize. I'd like to think Einstein would have endorsed this book as well. Nobel Prize committee, if you are reading this, my contact details are on the book sleeve.

Matter is all the *stuff* in the universe. The stuff that atoms make. That includes you and me. According to de Broglie, you too behave like a wave. So deep! So what is your frequency? By the way, don't ask this within the earshot of your kombucha-drinking acroyoga friend, for example, unless you want to also hear about their astrological sign and healing crystals. Us, though, we can do the math.

If you recall, h is a really small number. You don't have to go back. I'll paste it right here for you: 0.00000000000000000000 000000000000006626. You'll notice that h is at the bottom of the previous equation. Dividing by a tiny number is the same as multiplying by a huge number. For photons—light particles—their energy is also really small, so we end up with frequencies that are a reasonable size. You, on the other hand, have lots of energy. In the fancy Old English we saw in the last chapter, your kinetic energy is *the product of the mass of a body and the square of its speed.*

Your mass and your weight are not the same things, but

luckily the scale in your bathroom does the calculation for you. Let's suppose you weigh 70 kilograms. At a speed of 1 meter per second, you're walking a bit slowly because you are reading this book or staring at your phone. Your frequency is then 5.3×10^{34} Hz. That's 53 decillion or 53 million billion billion billion hertz. It doesn't even make sense to think about how big this number is without some context.

Suppose, instead of light or tiny particles, we wanted to perform a double-slit experiment with people. We see people going through doorways all the time, but we don't see an interference pattern on the other side. What would it take to see it? Well, first of all, the people would have to be moving imperceptibly slowly. Like, so slowly it would take the age of the universe to pass through the slits. That's fine. It's not like we have anything better to do. But also, the size of the experiment needs to be so large that the people would have to travel the length of the universe to resolve the pattern. I order my groceries online. If I can't even be bothered to walk to the store, I'm definitely not walking to the end of the universe. Besides, I've already been to Cleveland, and that place feels like the end.

The point is, we will never see this *wave-particle duality* effect at the scale of visible objects—not because it doesn't exist but because it is an impossibly tiny effect. Lucky for photons and electrons, though, because they are just tiny enough for us to detect *and* display their quantum beauty. In other words, you *do* have a quantum frequency, as does everything else in the universe, but *your* quantum frequency is so irrelevant that you don't need to worry about the efficacy of claims about things resonating with it. Irrelevance, however, is not a strong enough deterrent for some people. Before we go debunking, though, let's summarize what we've learned so far by making it bold.

> In classical physics, there are waves and particles, but they are separate things. In quantum physics, everything displays both wave-like and particle-like behavior.

You Almost Forgot the Crystals

Crystals. Am I right? It would seem all you have to do to dazzle people these days is mention crystals. Before I shit all over crystals, though, I'll grant you this—they do look cool. But here is what they don't do—they don't vibrate on their own and resonate with any part of your physical body or made-up shit like your chakras or auras or whatever ephemeral thing you want to imagine you possess. When people talk about crystals for healing and other bullshit, they are usually talking about gemstones. These are simple. They are polished minerals pulled from the Earth—rocks made shiny. Full stop.

In addition to being able to buy over one hundred thousand crystal-imbued products on Amazon—which offends me on many levels—there are over fifty thousand books about crystals. That's right, books. BOOKS. Like, the things that are supposed to contain facts and knowledge. Not only does *The Crystal Bible* exist, but *The Crystal Bible 2* is also a thing. I know what you are thinking—what the fuck needs to go in *The Crystal Bible 2* that isn't in *The Crystal Bible 1*? New crystal prophets pop up all the time, but it's not like people have invented new crystals. We've dug up and described in great scientific detail every kind of rock already—ever heard of geology? Wait...this just in, we *have* invented new crystals...my bad. According to common wisdom and lore in the crystal community, any individual rock you pick can be named and trademarked. No joke. New Zealand Dragon Eggs™ are just plain rocks with a purple tinge to them. But slap a sexy name on them, and you can convince the gullible that the vibrational power is so high that it resonates with the third-eye chakra, which sounds pretty good. Just what I was looking for. With these new healing abilities, I'll be able to grow back the fistful of hair I just pulled out.

And just when you think you've covered everything from amethyst to a polished turd stone, surprise! *The Crystal Bible 3* is also available to throw your hard-earned money at. With 250 "new generation, high-vibration stones for healing and transformation,"[ii] how could you pass this up? It's actually quite obvious how it works—if the crystal has high vibration, it will resonate with you, and you will have high vibration. The higher the vibration, the better, duh. The highest vibrations can even allow you to make contact with spirits in other dimensions, and, here's the thing, why the fuck am I telling you this? It's all obviously bullshit. Like, it's comically bullshit.

There are really only two types of crystal peddlers. The first are the straight-up grifters looking to make a quick buck who are willing to lie, cheat, and steal to get it. These ones will tell you that a stone's vibration will cure all your maladies. The other type are the big brands, and the thing about big brands is they have big lawyers, and big lawyers are smart—at least when it comes to defending the financial interests of shitty people. If you are buying healing crystals from a brand, I want you to take a look at the disclaimer in the fine print. That will tell you everything you need to know about crystals.

Come, Resonate with Me

"When you use the power of the quantum field, honing your emotions and intentions to create the frequency or vibration of that desire, you will draw the things you want to you."[iii] Uhhhh... what? That's from the book *Quantum Love* by Dr. Laura Berman. Not that that matters. Go to the New Age section of any bookstore, and you'll find volumes filled with this kind of shit. Let's—for the moment, just for the sake of argument—pretend that this sentence is not a random string of garbage words strung together in a semantically correct way without conveying any meaning. The gist of this sentence, and the countless others like it, is the following idea: things have a frequency, and if you match that frequency, you will resonate with it.

The first problem is the more obvious one. Real physical things like cash or a love interest do have a frequency, as we just learned from de Broglie. However, we also just learned that it would be impossible to display their wave-like properties in this universe. Besides, if I somehow move at the exact speed so that my frequency is the same as a pile of cash, it's not like a pile of cash is going to just magically fucking appear. So our charlatans must not be talking about the frequency of physical things.

But now we are even deeper into the bullshit, because the second problem is that abstract ideas—like love, desire, bravery, creativity, pleasure, etc.—do not have a frequency, quantum or otherwise. So far as I am aware, no serious attempt has ever been made to define such a concept. Social scientists attempt to measure feelings such as anxiety and other abstract notions such as intelligence, but these are always flawed. You can tell when people are not anxious or really anxious, but you can't measure anxiety in any precise or meaningful way. So first we would need a rigorous and universal way to measure these things, and only then could we come up with a theory for them, which even *then* may not use the concept of waves.

Unless some scientific instrument company is selling you modern equipment to perform the double-slit experiment, quantum frequency is not relevant to any aspect of your life, love, wealth, or happiness. Quantum love? Quantum bullshit.

One of These Waves Is Not Like the Others

Are all waves the same? Well, obviously no. How about this: are all waves with the same amplitude, speed, and frequency the same? Hmm...not so clear now. I haven't told you this explicitly yet, but no, they aren't. Some waves, like sound or earthquakes, move through air or earth, compressing and contracting the stuff they pass through. Other waves, like radio waves, are oscillating electric and magnetic fields that can travel through completely empty space. There is no reason these different types of waves should interact or *resonate* with each other at all.

For example, a sound wave at your de Broglie frequency of 53 decillion Hz would have to have a wavelength a billion billion times smaller than an air molecule. Not possible. Doesn't even make sense. When it comes to waves then—like real, actual waves and not bullshit quantum healing waves—it's important to understand what physical thing is oscillating. Sometimes it's not obvious, but 100 percent of the time, it is not the de Broglie wave of you being thought of as a massive quantum object. There is usually a wave, yes, but it's not quantum.

Can you resonate with anything at all then? Well, that depends on what *you* are. Your eardrums can resonate with sound waves at many frequencies. The photoreceptors in your eye have energy levels tuned to be resonant with photons of visible frequencies. Your whole body itself can even act as an antenna. Yikes. Waves are starting to sound scary!

All These Waves Are Making Me Sick

Cymophobia is the abnormal fear of waves, and there seems to be a lot of it going around. The sound that windmills make? Scary! Wi-Fi signals? Poisonous! Giant ocean swells? Those are actually dangerous. Don't fuck with the ocean. Joking aside, you wouldn't want wind turbine syndrome or electromagnetic hypersensitivity, would you? WOULD YOU? I didn't think so. By the way, those are real terms people use for fake diseases. Cymophobia is actually a real disorder though, so don't go telling cymophobics that waves are literally everywhere at all times.

Wind turbine syndrome, like all things nonscientific, is a catchall term for a bunch of gripes people have because they don't like the look of wind farms...or clean air. Quite a few scientific studies have confirmed that proximity to a wind farm does not increase the risk of any diseases. One study did note that participants routinely reported "annoyance" with wind turbines. At least annoyance is something I can sympathize with.

Electromagnetic hypersensitivity is a term used to refer to people who claim exposure to electromagnetic radiation (the stuff currently being emitted by the phone a few inches from your reproductive organs). Again, there is no scientific evidence here. In fact, people with strong claims about the sensitivity could not even detect the presence of the same radiation in well-controlled experiments. Checkmate, conspiracy theorists!

The interesting thing about claims against large organizations, such as corporations and governments that build wind farms and cellphone towers, is that they are taken quite seriously. Again, lawyers seem to be good fo... Never mind. I don't want to be quoted on that. One of the consequences of this is that we have actually figured out how to "cure" someone of a fake disease—many hours of expensive psychotherapy! Good thing mental health is taken so seriously in most societies... Can I put sarcastic-looking sad face emojis in here or what?

Technological progress is a double-edged sword. On the one hand, it can offer unmistakable positive advances, such as the mass production of vaccines. On the other, it can cause global destruction—such as the threat of nuclear war. Then we have things like the internet, which I'm not so sure about. I mean, have you spent more than five minutes on TikTok?

People are rightfully confused, especially when the media gets involved. Indeed, when it comes to claims about energy-harvesting wind turbines causing cancer or poisonous Wi-Fi, confusion correlates quite well with media attention. Scientific studies have shown that the claims are baseless, of course, but a little understanding of waves would give you more comfort.

According to a recent USA *Today* article, "Critics have

linked wind turbine operations to electromagnetic fields (EMF), shadow flicker, audible noise, low-frequency noise, and infrasound."[iv] These are all different types of waves. But we know already that we are surrounded by waves. Imagine that you are being bombarded with every frequency of every type of wave— sound, electromagnetic, gravitational, fuck it, even quantum love waves. That's not the important part. The important part is *how intense* each frequency is. Yes, windmills cause sound, but so does traffic. Just as you wouldn't want to live under a freeway, you wouldn't want to live under a windmill. It's annoying, but so are people who share their Goop success stories on Facebook. Wind turbine syndrome is not a medical condition, and windmills are not dangerous...unless you are a bird. Then I agree—fuck windmills. And planes. And windows. And cats. And garbage bin lids. And Wi-Fi—no bird wants electromagnetic hypersensitivity.

So what about Wi-Fi anyway? That's a bit closer to the quantum domain, what with it being made of photons and all. Surely there is a mysterious quantum reason we're not aware of yet as to why Wi-Fi is poisonous, right? And we are putting it in schools too! Think of the children! Oh, I do think of the children, the poor children who are exposed to these conspiracy theory blowhard parents and politicians.

Wi-Fi consists of radio frequency waves. Your phone is like one of those old radios we talked about, but it is tuned to the five-billion-hertz frequency of your internet router. The circuitry in your phone is very sensitive to these particular radio frequencies, which allows it to receive important information about how many people liked your latest tweet. Your body is also *tuned*, in some limited sense, to radio waves. Most radio waves pass right through you or bend around you. But your body will absorb the energy of some frequencies. That can sound scary, but calm the fuck down and listen to the scientists.

First, let's be crystal clear about one thing here: it is not as if scientists just ignore the potential threats of technology. The opposite of a technophobe is a technophile, and scientists are often portrayed in popular culture as spectacled lab rats recklessly pursuing science for the sake of science. But that's bullshit. Besides the medical studies debunking conspiracies and self-diagnosed maladies, scientists have also determined the actual measurable response of the human body to various waves. The resonant frequency of humans to sound is about 10 Hz, and the resonant frequency to electromagnetic waves is about 70 million Hz.[v] Health and safety agencies use this information to make decisions about how technology can be used. The Federal Communications Commission (FCC), for example, greatly restricts radio frequencies in the range most harmful to humans.[5] Spoiler alert: your body's peak absorption is not the frequency of Wi-Fi signals.

What would happen if you were exposed to the harmful waves? Nothing quantum—that's for sure. Remember, an interaction at the quantum level is between two individual particles, and the photons carrying the radio waves don't have enough energy to do anything interesting down there. If you want to know what radio waves *can* do to biological matter, you need look no further than your microwave oven. The energy absorbed turns into motion of the molecules of the thing being

---

5 Read much, much more at the FCC Radio Frequency Safety page: https://www.fcc.gov/general/radio-frequency-safety-0.

exposed to the waves. We have a name for the motion of smaller things that make up bigger things: temperature. So what some radio waves *can* do to you is heat up your tissue. There is nothing special about a burn due to radio waves—it'd feel no different than putting your hand on a hot stove. In fact, diathermy is a medical treatment using radio waves (and ultrasound) to heat up tissue. Like most technology, though, there is a fine balance between physically helpful and physically harmful, and a shitload of people in between who are just plain mentally harmful.

It's nice to think that plenty of research has been done on this stuff, specifically for the purpose of health and safety. Scientists, medical professionals, and, yeah, even the lawyers sometimes have figured out pretty well what harms electromagnetic waves pose and have recommended policies to prevent them. Engineers, on the other hand, are more interested in humans interfering with their precious Wi-Fi signals. In a confined place like an airplane, people damage radio signals a lot more than the other way around. So next time the Wi-Fi doesn't work on your way to that important business meeting, blame the person next to you for blocking all your internet waves.

So in the end, I guess I have to admit that you can say in very controlled scenarios that electromagnetic waves have vibrational energy that can be transferred to you. The problem still remains though—if you buy the *real* thing, you'll only get your ass burned, and if you buy the bullshit, well, you'll still get your ass burned. The energy in some radio frequency waves can heat stuff up. Sorry—no auras, no chakras, no heightened mental states or meditative enhancements. Also, no quantum. If you really wanted to damage your body on the quantum level with photons, you'll need higher frequencies, as we discussed in the previous chapter.

Wow, This Quantum Stuff Is Really Sounding Irrelevant

Now you are getting it! Quantum waves are just not something you can personally control to influence your outcomes. Sorry. No science fiction fantasies for you. If someone is selling or scaring you with quantum waves, they are full of shit. If you find yourself on Amazon browsing products with the words *quantum* and *resonance* in any combination, delete your account.

I don't want to leave you with the impression that quantum waves are entirely useless because they are indeed important for many technologies. It's just that, again, these are built up slowly and carefully through international collaborations of scientists that don't need to sell their ideas to you with sketchy links from horribly produced YouTube videos.

Take, for example, an *electron microscope*. First, let's recall how normal, everyday microscopes work. You have the eyepiece and the tube thingy that has a bunch of magnifying lenses in it and then the sample of bacteria or whatever, and finally, below all that, is a bright light. The light hits the bacteria, and the denser parts block more of the light, and voilà! You get to see tiny shit. But what is the smallest thing your microscope can see? That depends on the frequency, or equivalently the wavelength, of the light. Since our eyes can only see visible light, we are limited by what violet light (the shortest wavelength or highest frequency visible light) can resolve. We could use light with a shorter wavelength, but then we would need a different sensor other than our eyes to detect it.

But wait, what is a wave with a really short wavelength? Dun dun duuuun! Matter! Quantum physics to the rescue. I think it is safe to say that no one would have thought of throwing matter, like baseballs, in the dark to try to take a "picture" of something. But when you consider the wave-like properties of matter, the principles of microscopy are the same as for light. So the electron microscope was born. The image shown on page 50 is from an electron microscope looking at pollen. If you weren't allergic before, you probably are now!

The double-slit experiment seems like an esoteric relic of the past, and it is, but the *modern* version of the same experiment goes by another name: *interferometer*. As the name suggests, it is a device exploiting the interference of waves. Interferometers come in many shapes and sizes, from ones that fit in the palm of your hand to ones that are several kilometers long. Indeed, the latter size was necessary to detect gravitational waves for the first time in 2015. It is an indispensable scientific instrument for precision measurement. But even here, it is only the wave-like properties of light that are being probed. The events of particles being detected one at a time are far from achievable at this scale. Eventually, such technology will reach the ultimate precision limits set by the laws of physics, and we will again be confronted with the quantum conundrum. At the moment, though, such precision is not necessary for consumer electronic devices. So you won't find the wave-particle duality problem being presented to you in your daily life. Phew.

The only quantum waves that can hurt you are the ones you create for yourself. So dig your toes in the sand, watch the waves roll in, listen to sound waves of your favorite song, bask in the electromagnetic waves from the sunset, and tune your mind to the quantum waves of the universe, letting that vibrational energy cleanse the negative toxins from your... [CLICK HERE TO LEARN MORE].

3

We have no fucking clue what is going on

In high school, I was forced to read *The Great Gatsby*. And when I say read, I mean watch. Book reports are torture for a budding mathematician. The book...err...movie, though, has informed the entirety of my understanding of the 1920s. Given that, it's not hard to believe that a 1920s socialite would be interested in the latest scientific findings from the other side of the world. How very exotic and exciting—knowledge of such intellectual achievements would surely lure a love interest at my next dinner party. But this seems at odds with what we might expect today. Today's Gatsbys have reality TV shows about attempting the least intellectual feats possible. What the hell happened?

For most of scientific history, intellectual pursuits were possible only for those rich enough to buy free time. Today, of course, we all have free time, as you, dear reader, are clearly evidencing. Yet now we just squander it away looking at pictures of ourselves with digital filters and watching epic dramas on Netflix, usually without chill. How very depressing (at least after the binge is over anyway). It's not all bad though. We don't have smallpox, and scientific knowledge is no longer reserved for the social elite. Progress.

So what was the sexy new thing in science that reached public perception anyway? It was, of course, quantum fucking physics! The first conceptual hallmark of quantum physics to reach popular culture was the *uncertainty principle*. Colloquially, the uncertainty principle has been understood as the indisputable fact that some things can never be known. But this doesn't sound so profound. In fact, it sounds a bit lame.

However, there is a sense in which *you*, more than anyone in history, should find that a limit on knowledge is empowering. Whereas waves and energy seem like properties of the world independent of us, uncertainty pertains to knowledge, and that means we are back at the center of it. And nothing delights humans—especially twenty-first-century humans—more than being the center of the universe.

For if there is a limit on knowledge, it must be invoked only after I choose to obtain that knowledge. So in a sense, I invoke the laws of the universe—I create the universe. Yeah, take that, eighth-grade teacher who said I would never amount to anything!

You Know It

Do you even know what not knowing is and how you know when you're not knowing it? Lots of things can never be known. Any question of the form "what if X happened instead of Y?" can never be answered with certainty. Like, *what if Hitler was never born* is nice to think about, but we'll never know what *would* have happened because it didn't happen. This type of counterfactual reasoning is what makes humans special, by the way. I mean, we do it all the time. *What if I said this instead of that? What if I took the earlier flight? What if I didn't eat those hot peppers?* Testing our hypothesis allows us to determine the causes of things, which is usually beneficial. However, this is NOT the kind of limit on knowledge we are talking about in quantum physics.

The limit on knowledge implied by the uncertainty principle is much more fundamental. Its essence, which we will explore in detail, is not about a limit on *human* knowledge—although it does imply that—but a statement that some things are not even definable in a way that makes them knowable in the first place. Phew. Mouthful. Let's let a politician explain it. They usually speak with clarity, right?

In 2002, Donald Rumsfeld said there were "known unknowns" and "unknown unknowns" when making a case for war. He was chided for political doublespeak—and this *is* a terrible justification for war. But he was onto something. There are things you know you don't fully understand. Like, I know you probably have clothes on—at least I hope you do—but I don't know what color they are. That's a known unknown. However, I am only drafting a book with this sentence now, on August 27, 2021. Will it be published? Will it have readers at all? Perhaps an entirely new medium for consuming books will exist, or perhaps someone else will publish a book on the exact topic tomorrow. These are all unknown unknowns, things *I* didn't know I could even lack knowledge of.

Though unknown unknowns seem quite abstract, by the very nature of the word *knowledge,* someone could—perhaps already does—know them. Quantum uncertainty, on the other hand, shows us that some things are not even unknown unknowns. They are more like known unknowables. So there are known knowns, known unknowns, and unknown unknowns. In quantum physics, there are also known unknowables, but the vast majority are unknowable unknowns. You know?

I Know That I Know Nothing… Well, One Thing

Before physics was philosophy, and Socrates laid the foundations of scientific inquiry when he said, "I know that I know nothing." For what else motivates us to seek out knowledge but an admitted lack of it? But by the time *quantum* physics came along, our arrogance had gotten the better of us. Two hundred years of successfully applying Newtonian mechanics gave us the false impression that the universe obeyed a simple set of rules. You tell me where something is and where it is going, along with all the forces acting on it, and I can solve Newton's equations to determine the entire future of that thing. Since all Britain did in the seventeenth century was fire cannonballs all over the place, this was also a practical skill.

But the philosophers did so much more with Newton's laws of motion. You see, if I knew the position, speed, and direction of not just one thing but *every particle in the universe* at one instant in time, I could predict all that would happen in the future and know all that had happened in the past. This is, of course, practically impossible, but someone—or some*thing*— could in principle know it. That meant that everything was predetermined. This idea became so entrenched that, for early twentieth-century physicists, the universe behaved according to simple laws and was *deterministic*. In other words, everything is predetermined by the laws of physics. This, of course, put a damper on the idea that humans have free will. But no matter. That was a problem for philosophers to argue about. We had calculations to do!

Then along came a spunky young German named Heisenberg, who—although known better on TV for cooking meth—made seminal contributions to modern physics. The story goes that he had a horrible bout of hay fever and secluded himself on a rock island where he had nothing better to do but think about quantum physics. And just like that, a true eureka moment—the Heisenberg uncertainty principle was born. In 1927, he wrote—well, according to some translator anyway—"the more precisely the position is determined, the less precisely the speed is known, and conversely."[i]

Position and speed of what, exactly? Heisenberg was thinking about subatomic particles—you know, electrons and shit. But we can imagine a funnier analogy.

How Do You Find Your Friend in the Dark with Only Baseballs?

Suppose you and a work colleague—let's call them a friend for the sake of argument—are in a dark room. You want to know where they are, but you can't move, and they are incapacitated. This is what happens when you don't meet your key performance indicators. Anyway, you do happen to have a whole bunch of baseballs. I'm sure you see where I'm going with this. What? No! Of course you shouldn't throw baseballs blindly in the dark, listening for the sound of a painful groan. Kidding. Do that. What else could "work friends" be for?

Eventually, you will find where your friend is, but they might have some bruises to show for it. You can't find out where something is with baseballs without affecting it—that is Heisenberg's uncertainty principle, loosely interpreted. Now let's take this analogy a little more toward particles. Instead of a friend, suppose you are looking for a volleyball in the dark by throwing baseballs. Again, eventually, you will hear the sound of a baseball hitting a volleyball. I've not done the experiment myself, but years of cartoon sound effects have taught me it probably would sound like *boing, wah wah.*

So you've found the volleyball, but now what? It's long gone. Hitting a volleyball with a baseball is going to make it move. You found out where it was, but now you don't know where it's going. *Why wouldn't I just turn the lights on?* you ask. Fair point. Let's do that. Aha! There are all the balls. And also someone covered in bruises.

We usually determine where something is simply by looking. But looking is just the same as throwing balls and seeing where they bounce. It's just that the balls are really, *really* tiny—they are photons. Replace baseballs with photons and volleyballs with electrons and you basically have Heisenberg's microscope, which was his original thought experiment, suggesting the impossibility of precisely measuring the location and speed of a particle. Photons aren't going to have much effect when bouncing off a volleyball, but they can send an electron reeling. We too don't notice the billions upon billions of photons impacting us every second, but an electron would.

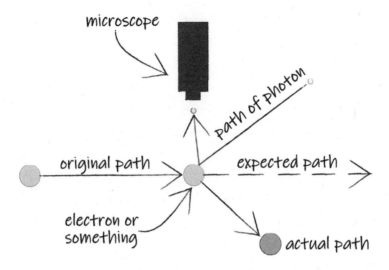

Perhaps, though, we just aren't clever enough. Perhaps there is a different, better way to locate things—something a little less invasive. Or maybe there is just an unavoidable problem with trying to make measurements of things. Even in that case, though, if we just never try to measure anything, could determinism be saved? In other words...

Maybe It's People Who Are the Problem

That's generally true, yes. But in this case, no, it's the universe. Even if no one is around to locate a particle or measure its speed, those two things can't even be defined for the particle in a way compatible with the rules of quantum physics, which we have tested to remarkable precision. The universe does not run like a mechanical clock but like the White Rabbit's pocket watch.

How does this uncertainty happen? You guessed it—waves. Okay, pop quiz. In the image on the next page do you see a wave or a particle? Trick question, right? There is certainly something there, and we can vaguely say where it is. But it is not a particle, that's for sure. And a wave? Maybe. It's wavy. But what is its wavelength? The crests seem to be different lengths apart. And the amplitude? Each crest has a different height. So it kind of has a location and it kind of looks like a wave, but it is neither. Wait a minute! We've seen something like this before—wave-particle duality!

Remember interference? That's when waves either add up at their crests or cancel when a crest meets a trough. By taking different waves and placing them on top of each other, we can make any pattern we like, so long as we have enough waves to work with. The more waves we add, the less it makes sense to assign the resulting shape a wavelength or other properties associated with a single wave. If we add more and more and more waves together, we can get the shape to look more and more like a lump in a single place. At that point, we would be very confident in assigning it a location—we might even call it a particle. Fuck it, why not? It's a particle for all practical purposes. But we would absolutely not assign it a wavelength—madness. Do you see the tension here? You either get something that is wave-like or particle-like but not both. Moreover, the more wave-like, the less particle-like, and vice versa. That's the Heisenberg uncertainty principle.

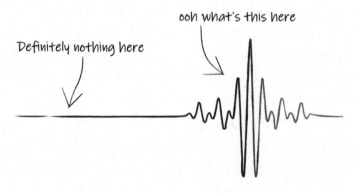

ooh what's this here

Definitely nothing here

That Feeling You're Having Is Called an Epiphany

I know what you are thinking: *he promised that the ground under my feet would remain stable, and now the foundation is crumbling beneath me!* That's right. Particles. Do. Not. Exist. Particles have a location, but the uncertainty principle already rules that out. So particles don't exist as we would like to imagine them.

Well, why the fuck do you keep using that word then? That's a completely fair question. I mean, you don't have to swear, but it's a valid question. Particles, particles, particles. We even have particle accelerators run by particle physicists. So if particles don't exist, what are these people doing? Probably drinking coffee, but that's not the point.

Particles are an idea, an idea that serves as an approximation to reality. Ideas that survive are useful. The Golden Rule is useful. It has survived. Dionysus, the god of the grape harvest, is dead. Neither are or ever were *real*. Particles are useful because of intellectual efficiency—they are simpler to think about than "a thing that kind of has a location and is a bit wavy but not specifically a wave per se."

Particles don't exist, just like the "average person" doesn't exist. But they are useful, just like the "average person" is a useful concept. We still talk about the average person even though such a person doesn't exist. More than that, though, parts of society are largely arranged for this nonexistent person! Teachers teach to the "average student," seats are sized for the "average ass," and this very book is written for the "average reader," who is definitely not you. You're way above average; don't worry.

So we talk about particles and even build multibillion-dollar experiments to detect them because it is a useful shorthand for an otherwise complicated situation. If that annoys you, if you feel like scientists have been lying to you your whole life, well, as I said earlier, dinosaurs had feathers—deal with it.

Waves, in this respect, are no different. They are the other side of the duality. Pure waves—that is, ideal waves with just one frequency—do not exist. The argument is exactly the same. For a wave to exist with a single frequency, it must spread out across all space, infinitely many evenly spaced lumps in all directions. That is to say, it would not have a location. Just like the approximation of a particle, the lump of waviness in the previous picture, we could have a different pattern that is an approximation of a wave. Starting with that picture, which can be thought of as the combination of lots and lots of perfect waves with different wavelengths, we can start to remove wavelengths. As we do, it becomes more plausible to assign the thing a wavelength, or at least a small range of wavelengths. However, this is at the expense of it becoming more and more spread out in location.

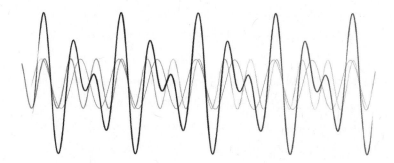

This view of uncertainty demonstrates that the properties of a thing are not so much *uncertain* as they are just not definable. An electron, for example, just does not have a unique and definite position. People—with their limited capacity to know stuff—are not necessary to evoke this fact. Yet there you are, poking around the atom.

You, Creator of Worlds

If you want to find a friend, a volleyball, or even an electron, you choose where to look. You might even be a clever engineer and intentionally *trap* an electron. In this scenario, you have taken an electron, which could have been anywhere, and confined it.

By choosing to measure the location of an electron precisely, you have forced the wavelength to become less defined and, hence, uncertain. Recall the connection, via de Broglie, between speed and wavelength. This is exactly what Heisenberg meant when he said, "the more precisely the position is determined, the less precisely the speed is known, and conversely."

Precisely determining the position of something is easy to intuit. When my children are adults, I'll have no clue where they are. When they are early adolescents, I'll have a rough idea. My young children are certainly in the house somewhere, and the infant is in the crib. The more confining the space, the more precisely the location is known. But what about speed or wavelength—what does a measurement of that look like?

Look again at the figure on page 62. How would you determine the wavelengths present in it? It looks rather complicated. But it is in fact the combination of only two wavelengths (shown as lighter dotted lines). One simple idea would be to start measuring the distance between peaks—that's the definition of a wavelength after all. The more peaks you include in your measurement, the more confident of the wavelengths you are. Ah, but now you've covered a lot of space, and you must wave bye-bye to the idea that things have a "location."

The very egotistical way of saying it is this: you choosing to measure one property of the world forces another property to be less defined. The act of measuring changes the system being measured. You choose what manifests in the world, or at least what part of the world gets fucked up. But I guess you already knew that.

Not Certain but Also Not Random Is Such a Random Thing to Say!

I know you are so clever, so you may be thinking ahead a bit. You may have heard that quantum physics is a *probabilistic* theory—that is, there is randomness built into it. The world, you might say, according to quantum physics, is fundamentally random. *Random* is often used to mean something that is opposed to *certain*. If something is certain, we know what will happen. If we don't know what will happen, we say that what does happen is, or was, *random*. So you'd be forgiven if you thought that quantum *un*certainty simply meant randomness.

Randomness is one of those words that seems easy to define but is anything but. Everyone has a good instinct for what is and is not random. For example, the outcome of a coin toss is random, right? Or is it? What if I was a jerk (I am) with a trick coin? Then the outcome would still be random to you but not to me. One person's random trash is another person's certain treasure! (Don't think I'm above tricking you out of your money to make a pedantically subtle point.) Let's try to distinguish this from something fundamentally random, something that no trickster could even be certain of.

Most—if not all—measurements that are made are random. For scientists and engineers, we call the effect *noise* but understand that there are "true" values for the measured quantities. For example, when you take a look at the photos on your phone the morning after, you notice that the pictures aren't as high quality as you remember from the night before when you were sure you were a professional photographer. At night, the photos turn out *grainy*—that's the noise. Your camera is "measuring" light, and there is just not enough of it in the pub. You could take two photos a mere instant apart, and the noise would look completely different. It's quintessential randomness. However, you know that that is a consequence of bad lighting and the poor decision to buy Apple products. The noise in the photo does not reflect the "true" image of the scene. In principle, with good enough technology, such noise will go away, and your nightlife can be as equally well curated on social media. That, or you double the opportunities to take pictures of your cat.

If we imagine continually improving the precision of our measurements—our cameras—we eventually run into a problem before perfection is reached. When you take a picture, you are attempting to capture something that you believe is already there. Even if that simply means just the position of an object, the quantum uncertainty principle tells us that what we seek to measure is not something predetermined. For example, we cannot capture an image of an electron at an instant in time. In attempting to push at this boundary, we force properties of the object to become undefined, and what we register as measured values become random. This is fundamental. There are plenty of quantum fraudsters out there, but none of them can predict the outcome of a quantum coin toss.

Of course, selfies of us and our cats are a long way from the size of electrons. As always, quantum physics applies to you, me, and cats, but the additional effects that quantum physics predict are not perceptible. I can't measure your location beyond the thirty-four digits of precision that Planck's constant proscribes, but I only need one or two digits of accuracy to hit you with a baseball. As far as mobile phones go, don't worry; there are plenty of incremental improvements to be had on those cameras, and new versions will continue to appear every damn fall. The engineers of the iPhone 40324 still won't be battling the quantum uncertainty principle. But when the time comes, you can remind them:

> In classical physics, everything is predetermined, and the precision in actual measurement is only limited by imperfect engineering. In quantum physics, some pairs of properties cannot simultaneously be defined, and attempting to measure one forces less precision in the other.

The Greatest Joke Ever Told

Wanna hear a joke? You may have heard this one before if you frequent the undergraduate physics club at your local college. No? Okay, here goes.

Heisenberg is driving when he gets pulled over. The officer asks, "Heisenberg, do you know how fast you were going?"

Heisenberg replies, "No, but I know exactly where I am."

The officer looks confused and says, "You were going 150 kilometers per hour!"

Heisenberg throws his arms up and cries, "Great! Now I'm lost!"

Hahaha! Gold. Pure gold—actually, silver. This is only the second funniest Heisenberg joke ever told. You'll have to read the next chapter for the real gold medal winner. In any case, you probably didn't know that a degree in physics opened up a whole 'nother world of comedy. There is so much stuff out there to laugh at—whether that was the intention or not.

But this joke is funny insomuch as you are willing to suspend belief in the laws of quantum physics because, of course, it can't work that way. In this case, we can actually do the math. Heisenberg's uncertainty principle in its original and precise form looks like this:

$$\Delta x \Delta p \geq \frac{h}{4\pi}$$

Ah, fuck! New symbols? Don't worry—it's not so bad. You already know h. Remember that was Planck's constant. The new symbol, Δ, the Greek letter delta, actually has nothing to do with quantum physics. So if you don't get it, it's not my goddamn fault. Δ is "error," which is the mathematical—or, more precisely, *statistical*—representation of uncertainty. This is science, so we can't just wave our hands around talking about vague shit all the time. At the end of the day, predictions need math. The error Δ is something you see more often than you probably think—unless you watch Fox News, where everything is stated with absolute certainty.[6] Error is usually heard in studies with randomly selected participants. For example, you might read in a newspaper *89 percent of people report not giving a shit about knowing math within a margin of error of plus or minus 3.1 percent, 19 times out of 20*. What this means precisely is... Oh, who am I kidding? You don't fucking care.

The point is, when you measure something, there is a bit of error. You'd think that error would go away with a better technique, better equipment, or less alcohol, but quantum physics forbids this. The previous equation shows that the errors when measuring the position of something (that's given the label Δx) and the momentum of something (labeled Δp) can't both be zero. In fact, the less the error in one, the more the error in the other! This is why Heisenberg says to the police officer—yeah, yeah, I'm explaining the joke—he doesn't know how fast he was going. He can't. He can't because he knows exactly where he is. If the uncertainty in his position is very small, the uncertainty in his speed (momentum) must be high. So when the officer tells him how fast he was going, the relationship flips, and Heisenberg must no longer know where he is. Funny, right? Go on, read it again. You'll chuckle.

Of course, this can't be relevant to real life. For one, you can't talk to police officers like that—unless you are a white dude, of course. But it should also be obvious that Heisenberg's

---

6 Every station cutting broadcasts to share "breaking news" every fifteen minutes is completely full of shit, in case you were wondering.

uncertainty principle doesn't play a role in everyday life; otherwise, someone would have noticed it much sooner. Let's quantify this like good scientists. Suppose the police officer's speed camera was accurate to 1 mile per hour. The average mass of a midsize car is 3360 pounds. Momentum is mass × speed, so Δp = 3360 pound-miles per hour in the equation on page 67.[7] Now divide Planck's constant by 4 times π times 3360, and we will have Heisenberg's minimum uncertainty in his position. Any calculator or even Google will tell you that Heisenberg's uncertainty in his position is three trillionths of a trillionth of a trillionth of an inch. I think it is safe to say he knows where he is.

The Heisenberg joke you might generously say is "inspired" by quantum physics, or at least a loose reading of it. However, if you thought lame physics jokes were the end of quantum bullshit, you've never met a "quantum-inspired" artist.

Art: The Most Obvious Application of Quantum Uncertainty

In 2008, Robert P. Crease wrote to the readers of *Physics World* asking for examples of quantum language in pop culture, suggesting "if we think scientifically rather than judgmentally, all uses of quantum language—whether precise or pretentious, technically correct or ill-informed and designed to impress— are interesting."[ii] Nope. Granted, Crease was hoping to find interesting, curious, or otherwise harmless examples. These

---

7 Don't know what a "pound-mile" is? Don't worry; I don't either. The important thing is that the units work out in the end of the calculation. In this case, if the position was calculated to have units of inches, a distance—it all checks out. Never doubt a mathematician.

cutesy examples, however, are buried deep in quantum bullshit. Judgment of the very scientific variety is warranted.

Six years later, by the way, Crease and Alfred Scharff Goldhaber did write a book with several examples.[iii] I'll let this excerpt speak for itself since I can't decide whether to laugh or cry or pull my last few hairs out: "So I decided to explore randomness and some of the principles of quantum mechanics, through poetry, using the medium of sheep."

Randomness is a common trope in artistic impressions of quantum physics. But in nearly every case of "randomness" you might encounter, quantum physics need not be invoked as an explanation. I don't want to say, for example, that quantum physics has *nothing* to do with why you keep losing at the craps table. Perhaps the casino does have some laboratory in its basement performing quantum coin tosses to generate all the random outcomes on the floor. More likely, though, the real reason you are losing at the casino is that gambling irresponsibly is only suitable for coked-up celebrities, and I don't think Gary Busey is reading this book.

More relatable popular culture examples of quantum trash are easier to come by. Take, for example, Professor Charlie Eppes, a fictional prodigy who solved mysteries with mathematics on the hit TV crime drama *Numb3rs*. While the show featured the keyword "quantum" many times, it was the very first episode (after the pilot) when the uncertainty principle enjoyed its fifteen seconds of fame.

Here's the script:

Charlie Eppes: There's something else that has to be considered.

Don Eppes: Like what?

Charlie Eppes: Heisenberg's uncertainty principle. Heisenberg noted that the act of observation will affect the observed; in other words, when you watch something, you change it, and, for example, an electron, you know, you can't really measure it without bumping into it in some small way. Any physical act of observation requires interaction with a form of energy, like light, and that will change the nature of the electron, its path of travel.

Don Eppes: Hold on. You know I got, like, a C in physics, so just take me through how this relates to the case.

Charlie Eppes: Don, you've observed the robbers. They know it. That will change their actions.[iv]

First, why is it that even fictional badasses like FBI Special Agent Don Eppes gloat about how bad they are at math and physics? Being innumerate should not be a point of pride, for fuck's sake! Besides, you shouldn't tell a mathematician to their face that you never liked math—that's like meeting Michael Jordan only to tell him that you never liked basketball.

Anyway, where were we? Oh, yes, Charlie Eppes. Actually, Professor, that's not how quantum physics works. Oddly enough, there *is* real science to what Eppes was talking about—just not physics. It's called the *observer effect*, and it's really quite obvious and intuitive. The observer effect, or the *Hawthorne effect*, is a characteristic reaction of people who know they are being observed. Unlike the scenario in *Numb3rs*, the effect is usually invoked as an explanation of behavior in sociological studies. I even evidenced this effect once.

As is the norm in most universities, study participants are university students who want ten-dollar gift cards in exchange for thirty minutes of answering questions about their feelings and shit. My task as an undergraduate was to investigate and install some software for a pretend task, then answer questions about why I chose the software. Of course, since I knew the researchers were watching, I actually slowly skimmed the end-user license agreement, pretending to read it, and remembered a few phrases from it. Then I was able to answer some questions about it correctly. But this actually screwed up their study! Why?

First, you'll be forgiven if you don't know what an end-user license agreement is since most of us, myself not excluded, completely ignore them. You see them when the windows pop on your computer screen, and you click "yes, yes, fuck off you pop-up, accept, accept, accept." Those are legal terms you are agreeing to. For example, when you install Instagram, you agree to give Instagram the same rights to your photos as you have, including the ability for them to sell your photos without even notifying you! But hey, fuck it. That's the cost of convenience and the opportunity to be an influencer, right? Anyway, the point of the study I participated in was secretly to observe how participants interacted with license agreements. But since I knew I was being watched, I acted differently than I normally would have. That's the observer effect in action, and I'm 100 percent certain it had nothing to do with quantum physics. That the robbers in *Numb3rs* were going to change their behavior after being observed also has nothing to do with quantum physics.

Not only is the psychological observer effect confused with the quantum one in popular culture, but even social scientists are also hopping on the quantum bandwagon express. Let's be clear: quantum physics has nothing to do with human behavior or feelings. The only human behavior quantum physics can possibly be blamed for is the delusions of grandeur of retired engineers.

Your Quantum Uncertainty Is Seeping into My Cultural Indecency

I'm all for quantum physics–inspired art, jokes, and shitty plot devices—cringeworthy or otherwise. However, the use of quantum language as a false authority is where I draw the line. The uncertainty principle is much more relatable, at least in its simple phrasing, than the mathematically defined quantum waves or esoteric distinction between continuous and discrete energy. That's why it appears so frequently outside scientific contexts. That also means we must be extra vigilant when we encounter it.

Recall that Heisenberg said, "the more precisely position is known, the less precisely momentum is known." We can always start here, appealing to authority, then subtly change the subject. Watch this. Think about the stock market. The current price is like the position. Where the price is going (up or down) is like the momentum. Therefore, the more you know about a stock price, the less you know about where it is going! Quantum finance or quantum bullshit? People are so easily trapped into thinking that because someone mentions quantum physics, they must be very smart, and their advice should be heeded. In fact, if someone mentions quantum physics outside a university lecture (or this book), you should kick them in the shins and run the other way.

And you must be extra careful when quantum bullshitters with actual credentials show up. Remember *Quantum Love*? It's not some obscure book I dug up in a thrift store. It has an over-four-star rating on the popular book review site Goodreads. If that wasn't enough, the author has an actual doctorate from New York University! So maybe dreaming about quantum physics while having sex *will* reignite the passionate fire in your relationship after all. But since Dr. Berman can speak from authority on quantum physics, then I can just as well give out relationship advice. Here it is: don't try to use quantum physics to save your marriage. Get a dog instead.

I don't mean to be picking on Dr. Berman as a nonphysicist here either. There are plenty of actual quantum physicists eager to spout off their bullshit. Since they won't be doing science when doing so, they'll often mask their ramblings as "philosophy." You can probably feel me shuddering through the page.

Quantum philosophy is big business. Remember the titles of popular quantum physics books from the list in the preface? Some of these physicists earn a lot of money spilling pretentious words and deep thoughts on pages in the hopes of being adored as public intellectuals. They'll tell you that the uncertainty principle proves whatever they personally want to believe about consciousness, free will, or whatever. It's enough to make actual professional philosophers cry.

Should one think deeply about these issues? Sure, why the hell not? But does that give you license to preach about morality and shove your armchair philosophical musings down everyone's throat? Fuck no. Some easily impressed sycophants buy it. But not us, dear reader, not us. We see right through the bullshit now. We're certified quantum cynics...err...I mean skeptics.

The Real Secret of Quantum Uncertainty—No Jokes This Time

Here we are again. This should start to sound familiar. Quantum phenomena, like the uncertainty principle, should not guide decisions in your everyday life. But that doesn't mean they aren't relevant to the science and technology that build our modern world. Be thankful you are being spared the boring details. The point of engineering, at least from the perspective of the user, is to create tools that are intuitive to use. Sure, we needed to understand quantum physics to create a laser pointer, for example, but would every lazy cat owner own one if they needed to solve the Schrödinger equation instead of just push a button? Doubt it.

Of course, modern physics and its engineering spin-offs are important for technological development, but surely the limited knowledge implied by the Heisenberg uncertainty principle is strictly a bad thing, right? Not so fast. Enter the secret world of cryptographers and hackers. Okay, to be fair, the academic world of cryptography is not as glamorous as Hollywood's portrayal of hackers. The cryptography expert in the office next to mine wears pleated slacks, not a Guy Fawkes mask, but he's still all about secret quantum stuff, so he's okay in our books.

Quantum cryptography is a straightforward application of the uncertainty principle. It works like this. If I want to send information to you, I need to use a telephone line or fiber-optic cable or something *physical*. A hacker, probably calling themselves Databurn or something equally lame, might be listening to the message. That's not good because what I'm trying to send you is probably important, since almost everything I have to say is important. But intercepting and reading a message written onto quantum objects, like electrons or atoms or whatever, is a *measurement*. Uh-oh! Because of the uncertainty principle, the hacker can't actually read the message without forcing a detectable change in the medium it is being transmitted in. Unlike existing technology, quantum physics makes communication perfectly secure!

The next time some bullshitter starts pontificating about quantum randomness, you now know what to do. No, don't run this time. Say, *aaaaactually, according to the quantum uncertainty principle, these things are not knowable—so you are full of shit!* Then kick them in the shins and run away.

4

That fucking zombie cat

If the Heisenberg uncertainty principle was the first concept in quantum physics to reach public perception, Schrödinger's cat is the most popular. That fucking zombie cat is everywhere, from poetry to television to blockbuster movies. Apparently, a cat being alive and dead at the same time serves as an excellent metaphor for life and even dating advice—though I guess it does kind of describe everyone I know using Tinder. But I'm going to impolitely disagree and call bullshit. Not a single incarnation of this godforsaken cat relates to quantum physics.

A cat

A fucking box

You're famous

I suppose at this point you're wondering who the hell Schrödinger is and what a cat has to do with physics. Well, I'm going to tell you. First of all, there is no cat. Forget about the poor cat. Schrödinger was real though. Erwin Schrödinger was an Austrian physicist turned biologist and one of the pioneers of quantum physics. Schrödinger's cat is a metaphor that he conjured up mostly to poke fun at his colleagues. However, it is now seen as an example of the most fundamental concept in quantum physics—superposition. Yeah, that's right. Super. Position. It's like position but super...er.

Like with everything else in quantum physics, though, the problem is that superposition inspires more bullshit than you could scoop out of a litter box in a hoarder's house. Superposition is what enables you to take two opposing things and say they can happen at the same time. This is a handy trick to have in your bag of bullshit if you are in the business of lying to people's faces after taking their money. But if you are an honest person... Bah, who am I kidding? Let's get crooked.

Did you know by reading your palm, I can bring our two consciousnesses into a *superposition*? Ooo, that sounds impressive! What does it mean? The hell if I know. But I could tell you that this heightened state of all possible awareness simultaneously allows me to see overlapping realities and reveal your future. Wait! I see...I see...a hospital. Yeah—you are there. Oh, no—you are in a gown, and you are talking to a doctor. She looks concerned... Ah, shit, it looks like your free five minutes are up. You'll have to pay to find out if you should have that mole removed.

No, of course that's not how superposition works. I do use superposition daily though. But I use it to perform calculations that solve engineering problems. Believe me—if I could use superposition to see the fucking future, I wouldn't need to beg people for five-star reviews on this book. Please though—I need those stars. I have mouths to feed! (Also, maybe some rum since you have me researching this quantum trash on the internet.)

Superposition Is Born

Do you remember the double-slit experiment? Chapter 2... about halfway through...wave-particle duality... Oh, come on! Fine, let me remind you then. Fire an electron gun—yeah, that's right—fire a fucking *electron gun* at a screen with two holes in it. Naturally, you expect to find two piles of electrons on the other side behind the holes. But that's not what you will find. Instead, you will find a pattern of many small electron piles in what looks like waves—an interference pattern—as if some electron fairy was trying to fuck with you.

Chapter 2 talked about the dual wave-like and particle-like behavior of electrons and other quantum shit. But let's back up a moment and pretend like we are in the early twentieth century. We are riding in style in horse-drawn carriages, wearing fancy handmade clothes, getting into fistfights and duels for fun, and... Oops, antibiotics didn't exist yet; now you're dead from diphtheria. I mean, do not for one instant think there was anything or anytime that can rightfully be called the "good old days." You thank whatever deity or random unforgiving universe you believe in right now that you live to bask in the glow of scientific progress. Sure, we've got problems (*cough*, climate change, *cough*), but we are better off now by any measure you can think of. Just look at those quantum physicists one hundred years ago—they didn't know shit!

One hundred years ago, physicists were trying to describe the behavior of particles, which they saw as tiny little balls. Electrons...err...balls ought to go through one hole *or* the other and fall in place on the other side. When it became apparent that this was not the case, physicists lost their shit. They argued endlessly in private and in public about who fucked up.

"You fucked up."

"No, you fucked up."

And so on, but in German probably. Then came the solution: *superposition.* Superposition was a significant advance in getting people to shut the fuck up. Superposition said that the electron didn't go through *either* hole but *both* holes at the same time. And instead of shaking their heads and saying, "What the fuck are you on about, Erwin?" they said, "Meh, good enough for us."

You Can't Handle the Truth

Okay, so that was a lie. That's not really what happened. But it is more or less how history has been rewritten. The reason is quite simple, really—what actually happened was, well, math.

Here is how science progresses. First, someone notices something curious. They are like, "What the fuck? I should tell someone." But then they forget about it. Because if they tried to tell other scientists, they'd be called a fucking quack. Time goes on, and more people start to notice the curiosity. When enough people notice, it becomes acceptable to talk about it. Other people then try to explain it within currently understood science. They fail. Next, some young upstarts begin to come up with new theories. Most of them fail too.

Meanwhile, new experimental techniques improve the

clarity of the anomaly, and more precise data becomes available. The theories become more refined and technical. Eventually, the mathematicians get annoyed enough to do something, get off their asses, and solve the problem by writing down some equations. Math for the win!

In quantum physics, it was no different. Experiments kept producing curiosities that could not be explained with the physics understood at the time. Plenty of scientists had theories to explain each new anomaly, but it was a bit of a shit show. While quantum physics was born with Planck in 1900, it was an ugly duckling for more than twenty-five years before it turned into the beautiful mathematical swan it still is today. In fact, there are *two* quantum theories: the *old* quantum theory and the *new* quantum theory. These are technical terms. I know, very creative.

Hindsight, they say, is 20/20. But that's bullshit. Hindsight is blinding. Once we understood how to think properly about atoms and elementary particles, we could no longer understand the perspective of those twentieth-century scientists who were confused about it. The world's greatest minds were struggling with something we now routinely teach to teenagers in the span of a few hours. It's impossible to grasp their collective confusion once you have internalized the modern perspective.

Back in the early twentieth century, physicists like Planck created what were seen as "corrections" to conventional physics. Most assumed that something would be found to seamlessly connect the quantum kludges back to Newton's physics. Of course, there were plenty of heated debates about how this should be done.

"You fucked up."

"No, you fucked up."

Still in German. Then along comes Paul Dirac, who rarely spoke at all. I like him already.

Revenge of the Nerds

Paul Dirac represents the stereotypical physicist you might see in today's blockbuster action/sci-fi movies or shitty sitcoms, except instead of talking too much, he more awkwardly said nothing at all. He is often referred to as a "genius," but at the same, his biography is titled *The Strangest Man*. Einstein once wrote of him, "I have trouble with Dirac. This balancing on the dizzying path between genius and madness is awful."[i]

Perhaps sometimes science needs a few interesting characters, but most geniuses are of the perfectly mediocre variety. (Ahem, maybe you know of one—winky face emoji.) Science is a collective effort among millions of people. We clearly cannot remember the names of all of them, and some are real assholes not worth naming, like Wade and Byrne—oops, too late. Stories need heroines and heroes, so we choose the most eccentric characters to remember. But they should be considered only symbols of scientific discovery and not celebrities to idolize and emulate. This brings us back to Dirac, who is perhaps worth emulating in some respects.

In the late 1920s, the field of quantum physics was at a crossroads. There were several competing theoretical frameworks for mounting experimental evidence that classical physics would be overthrown. This was significant. Classical physics reigned supreme for hundreds of years. The quantum upheaval was like a military coup, except instead of an army of soldiers led by a Napoleon, it was an army of fucking nerds led by the early twentieth-century version of Sheldon Cooper.

What Dirac did was show that *all* the competing theories were just different ways of looking at the same abstract mathematics. He called them "pictures" of a general theory he laid out in his book *The Principles of Quantum Mechanics*, the first textbook constructing all of quantum physics from scratch. It is here where quantum superposition was introduced: "The state must be regarded as the result of a kind of superposition of the two or more new states, in a way that cannot be conceived on classical ideas. Any state may be considered the result of a superposition of two or more other states, indeed in an infinite number of ways. Conversely, any two or more states may be superposed to give a new state."[ii]

Ummm...excuse me, what?

Superposition, Superimposition, and Supper Indigestion

Okay, before unpacking what this nerd just said, we need to talk about the word *superposition*. I need a good rant.

Why, Dirac? Why that word, superposition, I mean? It's a fucking terrible word to have used here. And no, it's not a poor translation. Dirac was English, and he was English when the English taught other English people proper fucking English. Dirac was taught the "Get this right, or I'll smack you with a lead pipe" kind of English. So why, Dirac, why? (Not sure if you guess this from the dates mentioned earlier, but he's dead and not going to answer.)

According to *Merriam-Webster*, which claims to have been defining shit since 1828, superposition means "the placement of one thing above or on top of another."[iii] As an example, the *law of superposition* is used. Note that this is a three-hundred-year-old principle in *geology* stating that young dirt sits on top of old dirt. I'll let you take a second to reread that. It's about as obvious as it sounds. Dig deeper, and you find shit that died a long time ago. That's the so-called law of superposition. Dirac chose this word for the new concept introduced by quantum physics.

Now consider the word super*im*position. *Merriam-Webster* claims this word, of similar age, is defined as the placement over or above something. The example is that of a triangle superimposed on an inverted triangle to form a six-pointed star. Nowadays, people are superimposing all the time on Zoom and Instagram by placing their beautiful faces above lovely backgrounds or placing stupid filters above their beautiful faces. Or, more often, it's superimposing images to make stupid memes about life clichés.

It is my contention that Dirac fucked up when naming the concept he called superposition, which it will forever be known by. Superimposition would have been a much better choice. Here's why. What is called superposition in quantum physics is the fact that adding solutions together to the equations of quantum physics produces new solutions. That is, if A and B are both solutions to some physics problem, then ½ A + ½ B is also a solution, and ½ A + ½ B is called a superposition.

Wrong! In what goddamn way is this similar to layers of dirt being placed on itself? We don't "stack" solutions together. We add fractions of them together—just like superimposing images! It's more like the saying *a little from column A, a little from column B*, and not like the saying *Oi! Dickhead, pile some shit from column A on top of shit from column B*. Okay, the last one is not so much a "saying" as it is something I hear people in high-visibility clothing say all the time.

One of the "fun" questions I am often asked by hipster millennials in their overproduced podcasts is the following: *If you could meet any scientist alive or dead, who would it be?* Haha, that's such a great question—wow, this *is* fun. I'll tell you who. It would be Paul Dirac. And since he isn't going to say anything, I'd start. *You fucked up, mate! Also, check this out. We call it toilet paper. I recommend you hoard some.*

State Your Purpose

Perhaps you thought this digression was meant to clear things up. But nah, I just needed to get that off my chest. I did, however, get us a little bit closer to superposition. Yeah, yeah, I know it should be super*im*position, but no one is going to die on this hill with me...unless... Oh, did...did you say something? No? Okay, never mind. As I was saying, superposition has one less syllable, so it's got that going for it.

Earlier, in my rant, I snuck in the phrase *solutions to equations*. That sounds a bit technical, and it is, but it's really what most physicists spend most of their time doing. The *problems* we face are phrased in mathematical language in terms of equations. You know some equations already from the previous chapters. But since I see you've already forgotten them, let me remind you of the one equation *everyone* knows, $E = mc^2$. Here, the letters are stand-ins for arbitrary numbers that represent energy (E), mass (m), and the speed of light (c), which is also squared (recall $c^2 = c \times c$). This is Einstein's famous equation saying that "mass and energy are the same things." Well, not quite the same, but related by the speed of light, squared. Now suppose you know the energy going into some reaction and want to predict how much mass could be produced. Well, you'd simply *solve* Einstein's equation. The answer would be the *solution to the equation*. Fuck yeah, you just did physics! (Not really, but we'll take the small victories.)

The best equations have solutions that tell us the *state*. In many of the sciences, including physics, the *state* of something is everything that can be known about it. We use the word *state* colloquially in much the same way. *Colloquially*. Do you like that? I have this word-a-day app that makes me sound really smart sometimes. Apparently, colloquially is also worth 76 points in Scrabble if your dog or children haven't eaten the only letter Q piece. Damn, another tangent. Actually, wait. Holy shit, this might work.

Suppose you are cool like me and my physics friends, and you are playing a board game. Scrabble is probably not your thing, so let's go with Monopoly, which you might also know by the name "sell me fucking Park Place, you piece of shit. Hey where are you going? The game's not over..." I may have a problem. Anyway, now imagine I flip the table. (Not in anger—it was an accident.) Could the game continue? Yes! Provided you knew where all the pieces were, you could set the board up exactly how it was before the accident. You knew the *state* of the game.

The state of physics stuff is the same—it's like a recipe to make something in a uniquely specific way. The state of a board game is easy to figure out just by looking at it. With quantum things like atoms, "looking" is more subtle due to the uncertainty principle, but the idea is the same. If you know the state of an atom, say, you know all there is to know about. In fact, like a sequence of moves in a very friendly game of Monopoly is a recipe to create a particular state of the game, a sequence of laboratory instructions can produce a unique state of an atom.

So now that we know what a state is, we have all the pieces to understand the quantum superposition puzzle—well, as much as we can understand from an eighteen-dollar book.

Six Impossible Things before Breakfast

Once you internalize quantum physics, it's the states of objects in classical physics that seem weird. They come from a very limited set of possibilities as prescribed by the laws of Newton. The Newtonian view of the world is very lame, like a movie that went straight to video with shitty actors you've never heard of. The view of the world Schrödinger gave us is like a big-budget blockbuster with Tom fucking Hanks.

Classical objects are those of the everyday world like people, animals, coffee cups, smartphones, and other shit—including literal shit. Quantum objects used to be things accessible only with big fucking microscopes and other expensive lab equipment. Now, quantum objects bend to our will and include atomic clocks, laser light, superconducting circuits, and wild and exotic things. Don't google that last one though.

The remarkable thing about states of quantum objects is that they can be understood at all. We speak about them in terms of the language of classical objects, which is also the language in which we communicate with one another. This is helpful but can also get us into trouble. It's probably best to consider an example.

Back before you or I were born, people would communicate by handwritten letters sent by mail. I know, charming. These might have been delivered by train. The train track is a classical object. The state of the train track is the location and speed of the train. Back and forth, there can be only one train on the track.

Nowadays, we communicate with memes and 280-character shit posts sent via pulses of light in fiber-optic cables. I know, charming. The path along which your pointless messages are sent is a quantum object. The state of the cable is the sum of the state of all the light it carries. The *sum*. The state is a superposition obtained by adding up the independent states of light pulses, which—because they're independent—can pass right by one another in opposite directions. Back in the classical world, we cannot create a superposition state of two trains going in opposite directions. That would be bad, though CNN would love exclusive coverage.

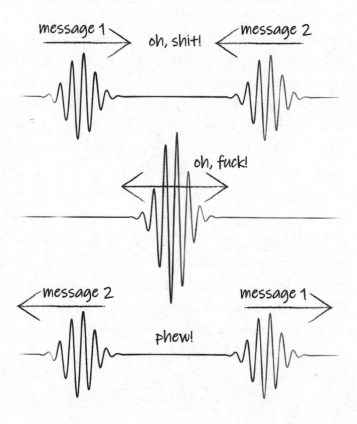

So classical objects just don't support the possibility of super-position. Time to raise a cat from the dead.

Is a Cat a Quantum Thing?

Light, by the way, is a wave of electromagnetism. So you can flip back to chapter 3 and the picture of two waves being added together to see superposition in action. Classical states are like the basic boring old waves we started with. Quantum waves include those *and* all the possible ways they can be added together. Quantum stuff just possesses a richer set of possibilities for its state of being. While you are made of quantum *stuff*—atoms and energy—sadly, you are not a quantum *thing*.

Superposition states are fragile. That's why they are not *obvious*. If they were, it wouldn't have taken so long to discover atoms and their structure. Once the intricacies of the quantum world are amplified up to human scales, they lose their richness. The bigger the object is, the more fragile any superpositions will be, which finally brings us back to the cat.

The Schrödinger cat paradox goes like this. Consider the following chain reaction. First, imagine there is a device that uses a radioactive decay event to trigger a hammer that breaks a vial of poison. Put this device in a sealed box with a cat, and suppose a radioactive decay event happens, on average, once per hour. Now, after thirty minutes, the chance of a radioactive event is 50 percent. Since the radioactive substance is a quantum thing, it is in a superposition of "not decayed" and "decayed." Thus, the hammer is in a superposition of "not triggered" and "triggered." The poison is in a superposition of "not released" and "released." And finally—don't tell PETA—the cat is in a superposition of "not dead" and "dead." Zombies weren't a thing back then. Otherwise, Schrödinger totally would have called it a zombie cat.

Schrödinger was showing how, in principle, quantum super-position could be amplified up to the size of a cat. But we never see zombie cats. I mean, they don't even put them in horror movies. What's the deal with that? I want to see a zombie cat movie. Hell, I'll even watch a straight-to-streaming movie about it. I'll even let you call it *Quantum Superposition*. Looking at you, Disney. But that would be the only place to see zombie cats, and this poses a problem. If the quantum world contains superpositions and the classical world does not, where exactly does the distinction lie?

Suffice it to say, we don't actually know how fundamental quantum processes eventually come together to create humans, thinking ones or otherwise. This is kind of the biggest leap of faith all practicing physicists believe in. I mean, it must be the case that somehow atoms interact in such a way that when seven octillion of them get together, they stop acting all quantum and start acting like fucking idiots.

This idea goes as far back as the "father of quantum theory" Niels Bohr vaguely stated it in 1920 in what he called the *correspondence principle*. The argument is succinctly stated thusly— I'm pretty sure that's a word. Big things obey classical physical laws. Big things are made of small things. Small things obey quantum physical laws. Therefore, quantum physical laws must contain within them the fact that many small things act together in a way that conforms to classical physics.

It is true that you and the cat are made of quantum stuff that obeys quantum rules that have superposition states available. But somewhere along the way, as we scale up from the tiny, invisible atom to large (no offense), highly visible you, quantum physics starts to mimic classical physics. Schrödinger and others a hundred years ago did not have the technology to probe exactly where this transition from quantum to classical was. Nowadays, we can see in highly controlled scenarios that this transition is blurry and variable, depending on many details of the experiment. But it still happens at scales far beyond our experience. So you, me, and the cat are doomed to remain in the classical world. It's probably for the best.

> In classical physics, objects have clear and well-defined states. In quantum physics, objects are mathematically described by adding fractions of classical states together.

The Greatest Joke Ever Retold

I have a joke. Maybe you've heard this one. It goes like this. Heisenberg is driving *with Schrödinger* when they get pulled over. The officer asks Heisenberg, "Do you know how fast you were going?"

Heisenberg replies, "No, but I know exactly where I am."

The officer looks confused and says, "You were going 150 kilometers per hour!"

Heisenberg throws his arms up and cries, "Great! Now I'm lost!"

The officer looks over the car and asks Schrödinger if they have anything in the trunk. "A cat," Schrödinger replies.

The officer opens the trunk and yells, "This cat is dead!"

Schrödinger angrily replies, "Well, it is now!"

Ha! What did I say? Now that is pure comedy gold. Of course, superposition by itself isn't that funny. I mean, I've never introduced any mathematics in a lecture that was received by an eruption of laughter—applause, yes, definitely, but not laughter. That cat though... Schrödinger inadvertently did more for the popularity of quantum physics with that cat than he did with the actual science he produced. Schrödinger's cat is apparently hilarious. In addition to appearing in any sitcom that features a "nerd" character, the cat is a favorite among comic artists. In 2013 alone, the popular *Dilbert* had three comic strips devoted to Schrödinger's cat. The punchline in every single instance is limited to "alive and dead at the same time." And if it isn't alive and dead at the same time, it's...

In Two Fucking Places at Once

Superposition is always misused in the same general way. Take two familiar concepts that are opposite and definitely cannot happen at the same time, and then say the following: *because of quantum superposition, these two things can happen at the same fucking time!* If you want to sound extra clever, call it a *paradox*. Here are a few examples I just made up because it is that cheap.

- I think I'm in love, but I'm not sure how she feels. It's like Schrödinger's relationship—love and not love at the same time.

- I feel like I can afford all the things I need but none of the things I want. I have Schrödinger's bank account—I'm rich and poor at the same time.
- I finished the book but didn't remember a thing about it. It's Schrödinger's book—read and not read at the same time.

Amazing, isn't it? I bet you are impressed...and unimpressed at the same time!

Remember Professor Charlie Eppes and *Numbers*...err...I mean *Numb3rs*? After fucking up the uncertainty principle, he's back to redeem himself with superposition.

Charlie Eppes: It's like the evidence proves him right and wrong at the same time.

Dr. Larry Fleinhardt: Oh yeah, the old paradox of Schrödinger's cat.

Alan Eppes: Is that that Persian that keeps hiding out in our garage?

Charlie Eppes: Uh-uh, it's an intellectual exercise.

Alan Eppes: I knew that.[iv]

What is this guy a professor of exactly? And who named this show? *Numb3rs* sounds like a username a thirteen-year-old Xbox player would use. I've never actually seen an episode, but I'm quite confident in my one-star review.

So right and wrong *at the same time*...sigh. We know the connection now. Schrödinger used the cat to illuminate the problem with the idea that there was no problem with making a sharp distinction between the world experienced by atoms and the world experienced by humans or cats. He "put" the cat into a superposition state, knowing all along it was ridiculous. He definitely was *not* demonstrating that anything you want can be in superposition. If anything could be alive and dead at the same time, it's Schrödinger rolling in his fucking grave.

So...uh...sure, Charlie Eppes, it's an intellectual exercise—as much an intellectual exercise as Shake Weights are physical exercise. But I'm not going to blame any of the writers of obvious fiction here. Most of it is in good fun. The problem here is that these bastardizations might be the only thing obsessive consumers of shitty television hear about quantum physics. You are lucky, though, because you picked up this book (or stole an online copy of it). And now you can either chuckle along or heavily roll your eyes when quantum superposition gets mentioned.

Enough of this low-stakes shit. What we really need to be cautious about is the rotting, stinking horseshit peddled by the quantum trickers. For that, we need to do some digging. Luckily, we have...

Dr. Chris Ferrie—Quantum Detective

In all my time working on quantum physics, I have never had such great admiration not only for a man's character but also for his work and humor. Having laboriously given my efforts to this theory for fifty-odd years, I am now narrating a new chapter with a new protagonist, Dr. Chris Ferrie. I have come to the point, however, where such a story...indeed, where such a narration has proved so strange. The truth, I must confess, is that I—Dr. Schrödinger—am dead. How this narration is working, I am sure you will be asking. Well, it has something to do with quantum physics, a cat, and probably too much rum. I believe that I have somehow been given this chance to follow Dr. Ferrie one final time—to follow him until he has, I hope, discovered the murderer of my dear cat, metaphorically of course. In any case, I said quantum physics, so it should all sound plausible to you. Now, to scribing the venture at hand!

Dr. Ferrie sat over his laptop, his right index finger hovering above the Enter key. He was always told not to talk to strangers on the internet, but now not only does he talk to strangers on the internet, he tells them where he lives, invites them to drive there, and then gets into their car. He is also extremely polite to the potential murderer lest they give a poor review, ruining his 4.9 star Uber rating. Thus, Dr. Ferrie was now immune to the dangers of the internet...almost. The question at hand was whether he would allow "can quantum superposition help my back pain" to enter into his search history. Facebook's algorithm would eat that shit up. But how else was he to find clues leading to the most villainous quantum bullshitters ever?

Dr. Ferrie teetered on the edge of the YouTube rabbit hole. What seemed like hours passed as the temptation to click on each recommended video tortured him—9/11 conspiracies, pizzagate, plandemic...on and on it went. Was the Earth truly flat? How long has the government been lying? Dr. Ferrie was discovering the truth, and all it took was doing his own research...albeit by reading the titles of asinine YouTube videos. The perseverance paid off when the algorithm started showing him alternative therapy suppression theories.

Quantum DNA healing... Hmm, *he thought to himself.* Now we are onto something.

Holy shit. I'm good at this narrative fiction stuff. What I'm apparently not good at is finding high-quality quantum bullshit on the internet. I mean, I even went looking for it. I was knee-deep in bullshit with quantum this and quantum that. I found books—like actual books someone decided to publish—on *Quantum DNA Healing* (Althea S. Hawk), *Quantum Success* (Christy Whitman), *Quantum Love* (Laura Berman), *Quantum Touch* (Richard Gordon), and even *Quantum Marketing* (Raja Rajamannar). But none of these even mention quantum superposition, arguably the most basic concept in quantum physics. *Quantum Theology* (Diarmuid O'Murchu) does mention it once in scare quotes. But beyond that, all these books employ the exact same tactic—couch some obvious and amateur advice in fancy language using the word *quantum* a few times.

There are plenty of examples promising at least some faithful analogy to quantum physics, but all fall way too fucking short—if they get off the ground at all. I suppose if you want feel-good, positively reinforced marketing advice but prefer it sounds all quantum sexy, fine—go for it. But don't be swayed into thinking something might work for you because of quantum physics.

If You Hate Something, Set It Free

You know what? I was ready to let it all go. I really was. Maybe quantum bullshit wasn't so bad, I thought. Then someone told me about fucking Goop, which—to the best of my knowledge—is about alternative medicine, Gwyneth Paltrow's vagina, and quantum physics. If you are interested in more than one of these things, I'm here to help. Well, maybe—it depends which one interests you most.

According to Wikipedia, "Goop is a wellness and lifestyle brand and company."[v] Gross. This is a combination of two of the worst inventions in human history—pseudoscience and brand marketing. Here's a good idea. Give them a Netflix special. Oh, and put the word *lab* in there to make it sound legitimate. Fuck me. Netflix executives must hate humanity. Not only could I not be bothered to spend fifteen minutes researching Goop, but I also couldn't bring myself to watch more than fifteen seconds of this godforsaken show. Luckily, my trusty YouTube algorithm—primed on hours of an absolute dumpster fire of unintelligible trash—recommended a few key clips of *The Goop Lab* to me. Here's the most offensive utterance from the show:

"There's been some amazing research done in Quantum Physics to support this. One foundational study is called the Double Slit experiment. Proved empirically without a shadow of a doubt that our consciousness actually shifts or alters—in some way, shape, or form—physical reality."[vi]

I'm sorry, what? I'd love to tell you what the fuck this bullshitter was talking about, but it really doesn't matter, does it? No, because this makes absolutely no fucking sense. I really can't bring myself to watch it again to find out for sure, but I think the show was about some "healer" who waves his hands over people as they spasm, apparently being exorcised of their last few brain cells. Actually, forget what I said before. You should go watch some clips of this episode to really get a sense of the sheer stupidity of it. *Of course*, quantum superposition—and certainly not some routine experiment performed by undergraduates—can't explain what is going on when the self-proclaimed healer telekinetically molests his inexplicably attractive patients. But I can. They are all completely self-indulgent assholes, oblivious to the fact that 90 percent of people know they are full of shit, and the other 10 percent don't give a fuck. No. No. No. There is no science in this at all, let alone quantum physics.

I'm not even that mad about Goop. What really annoys me is how and how much people react to it. The worst culprits come from the "skeptics" community. These people love to play the part of the infinitely rational scientist, keen to follow the evidence wherever it may lead. Here's the thing: science is guided partly by logic, partly by intuition, and partly by luck. If we had to evaluate, test, or replicate every stupid claim we heard, nothing would ever get done.

Next time you get a whiff of something that smells like bullshit, don't pick it up just to be sure. You have a license to call it bullshit, turn, and walk away. Don't forget to run if you ended up kicking anyone in the shins.

5

Faster than fucking light

Forget it. I give up. Fuck this chapter.

5

Faster than fucking light

No. I can't quit now. What kind of example would I be setting? So what if staring at this bullshit all day is giving me an ulcer, right? My chiropractor said he could work it out anyway, so let's forge on.

Love. It's enough to make you sick. It's been said that no other subject has had so many words written about it. Who said that? Who cares. It's mostly bullshit anyway, and if there is anything I've taught you, it's that bullshit should be completely and wholeheartedly filtered beyond any... Ah fuck, you stopped paying attention and started thinking about your favorite lame quotes about love, haven't you? Fuck it. Fine. Let's do this.

Love. Love is our connection to the cosmos. Love is entangled particles forged in the heart of stars waiting billions of years to ignite its quantum resonance. To truly experience love, you must enter the energy field of this romantic entanglement. Now, I know what you are thinking: *Chris, this sounds like that bullshit you warned us about*. But don't worry. It's all explained in my MasterClass on love. Sign up now! Fair warning though—it's mostly just a tutorial on how to sign up for a Pornhub premium account.

Now, *entanglement*. That's a new word. Sounds *spooky*, doesn't it? Oh...no, not really? Well, Einstein called it spooky. But I guess you know better than him. Hmm...maybe you are onto something. Yeah, come to think of it, I agree with you. Fuck that Einstein guy. What did he know anyway?

I can see you are a bit confused. That's fair. This chapter is not getting off to a great start. I probably should not have tried to write this drunk while this airplane was taking off. I'm tired. Is anyone else tired?

Technobabble

Unlike energy or waves, entanglement is not something you hear about every day—unless, of course, you are a customer service representative at a garden hose supplier. "Hi, yes, it happened again. The fucking hose is all entangled! My azaleas need water, damn it! They just don't make things like they used to. How do I unentangle this piece of shi..."

Entanglement. It's a terrible name, actually. It has nothing to do with things being *tangled*. There is not some "quantum" version of a garden hose being twisted up and knotted, so get that picture out of your mind right now. A better word would be *quantum correlation*. Oops, that's two words. Okay, naming things is hard. But the point is, the everyday analogy this quantum concept is all about is *correlation* and *not* some mystical entwining between the hearts of star-crossed lovers.

So entanglement is quantum correlation. Actually, I don't like the word *correlation* either. I can't quite put my finger on it. It just *feels* too technical sounding for a very simple concept. The synonyms the thesaurus offers me are even worse—association, interrelationship, interdependence, correspondence, concurrence, holy shit. Not that last one; "holy shit" wasn't listed—but holy shit. There doesn't seem to be a simple word relatable to a four-year-old for a concept every four-year-old understands.

Energy is a much more difficult to define concept. Yet everyone—even a four-year-old—has an intuitive understanding of it. You see the word *energy* and think, *yeah, I'm pretty sure I know what that is*. We talk about it coming from the Sun or our food. We talk about having it or needing it. It's probably in the name of that fucking company always sending you outrageous bills in the mail. I asked for paperless billing—why are you still sending me mail in 2023?! But now try to write down a definition of energy. Not so easy. To save you from getting up, I'll write down what you would have seen had you googled it:

"Energy is the quantitative property that must be transferred to an object in order to perform work on, or to heat, the object."

Look at that again. We define energy as *the thing* that's required to do *something else*. It's almost an admission that we don't really know *what* energy is. Yet we use the word—more or less correctly—all the time. Correlation is the exact opposite of this. It's super easy to define—a relationship or connection between two or more things—but we rarely use it in casual conversation.

When we see a pattern of things happening at the same time or one after another, we say they are *correlated*. Clouds and rain. Fire and smoke. Russians and vodka. Me and vodka. Me and Russians. Wait, no—got carried away there. Not that I don't like Russians—they just never arrived in the mail. But the rest of these are all correlated. They are also somewhat obvious. Subtle correlations, on the other hand, require careful observations and controlled experiments. Such are the diversions of scientists.

Nearly every piece of science news attention-seeking editors think is worthy of our eyes is about some newly found correlation. Chocolate and happiness. Chocolate and weight loss. Chocolate and heart disease. Hell, there is even peer-reviewed scientific research about a correlation between chocolate consumption in a country and the number of Nobel Prizes its citizens have won.[i] And this is just chocolate. You can repeat all the same bullshit studies and find correlations between Nobel Prizes and just about anything. Though you might find a negative correlation between Vegemite and Nobel Prizes.

Repeat after Me, *Correlation Is Not Causation*

I think every science journalist has to have this branded on their ass before finishing their degree. You could ask them about it—they like to show it off.

In the science section of the news, they don't use the word *correlation*. We hear things like eating sugar is *linked to* cancer, but this is often confused to mean eating sugar *causes* cancer. These are two completely different things. The researchers doing these studies have only some observations of people who eat a lot of sugar and also have cancer. Maybe eating sugar causes cancer, but maybe having cancer causes people to eat more sugar. Or, more likely, something else—like just being generally unhealthy—leads to both eating too much sugar and getting cancer.

In all cases of correlation, our minds crave a causal connection. Like that four-year-old child again, we want to know why. Since we only see the two events, we default to the option that one caused the other. But in most cases of correlation, it is actually an unobserved event that caused both of the correlated events. A famous example is the fact that a city with more police has more crime. Does more crime lead to more police? Do more police cause more crime? Actually, neither. In fact, a city with a large population has both more police and more crime simply because there are more people, period. This is called a common cause, and it's very important. Pay very close attention to that fucking man behind the curtain—he's the cause of everything.

The reason why very few scientific studies can say one thing causes another is that all possible common causes need to be ruled out, and there are infinitely many potentially hidden common causes. So to *prove* that sugar causes cancer would take years of work from scientists and medical practitioners. Recall how long it took to solidify the claim that smoking causes lung cancer, and scientific articles are still being written about smoking! The Google Scholar scientific database has recorded over four million (!) scientific works about the health effects of smoking. Meanwhile, there are only a dozen or so follow-up studies about the chocolate–Nobel Prize debate. So the jury will be out on that one for a while. I'm sure the *New York Times* is waiting with bated breath.

Cause for Concern

Cause and effect are basically the framework we use to navigate the world—*if* I do this, *then* that will happen. If I eat, then I won't be hungry. If I shower, then I won't be smelly. If I eat in the shower, then…well, I don't know. But the only reason to consider the idea further is to understand the cause–effect relationship. For science! Be right back. No. Turns out it was a bad idea.

Causes that are obvious are carried out by your body subconsciously. Some you do from birth without being taught, like breathing to get oxygen or blinking to keep your eyes moist. Other causes you must first learn, but then you carry them out subconsciously, like the subtle motions needed to balance on a bicycle. So far, these examples have all been *deterministic*. If I take a breath, then I will *definitely* get oxygen. (If that is not the case for you, I don't even want to know what fucked-up shit you get up to.)

Nondeterministic causes are more difficult to understand. They are often subtle, even if they are not portrayed that way. Smoking, for example, does *not* generate cancer cells with every puff. Hell, there are 102-year-old smokers who are about as healthy as a 102-year-old could be. (Not that healthy, generally. They could drop any minute.) Smoking merely *increases the chances* of getting cancer. That doesn't sound like the typical use of the word *cause*, but put it this way. What could you do to give someone lung cancer? Well, beyond implanting cancerous cells into their lungs, the only thing that comes to mind is forcing them to smoke. Not that scientists think in exactly this perverse of terms, but this is the sense in which we say *smoking causes cancer*.

Speaking of cancer, let's get back to quantum physics.

What Caused Entanglement?

Entanglement is a kind of correlation between quantum things that is super subtle. Of course, it has always been present in fundamental particles—we just didn't notice it. It also has been present in the mathematics of quantum theory since the early twentieth century, but it took Albert Einstein decades to finally find it. It might seem odd to say he "found" it, as if it were intentionally hidden somehow. But when the theory and equations of quantum physics were written down, not all their implications were obvious. Einstein was a quantum sleuth—he went looking, exploring with calculations and logical arguments, for consequences of the original theory that others hadn't seen yet.

Einstein was famously annoyed with quantum physics. He spent much of his academic life trying to find its weakness. If I were to toe the party line on this topic, I'd have to say Einstein fucking hated quantum physics because he didn't understand it.[8] But I'm not going to do that—I actually admire Einstein's attitude toward quantum physics. That's saying a lot too, because elsewhere, I've given advice not to have heroes—they will always let you down. Interestingly, much of what we know about Einstein's opinions on many things are from private letters to colleagues, friends, and family. The most infamous thing ever attributed to him—"God does not play dice"—was written in a personal letter to a colleague. We judge him by many of the views expressed privately in these letters, many of which were shitty views not worth repeating. To put that in perspective, that's like judging someone today by the content of their privately shared Facebook posts. Yikes. Please, please never judge me by the content of the tweets sitting in my drafts folder.

Many things written about Einstein and quantum physics paint a picture of him as a kind of Don Quixote tilting at quantum

---

8 As one science writer put it, "Einstein hated the quantum sandbox, especially the part about entanglement." Jeffrey Kluger, "What Einstein Got Wrong about the Speed of Light," *Time*, October 22, 2015, https://time.com/4083823/einstein-entanglement-quantum/.

windmills. But there is a meta lesson here that all these assholes can't seem to grasp about Einstein. It's not so much that Einstein wanted to tear down quantum physics because he didn't understand it. It's that Einstein wanted to peel apart the layers to get to a core he could understand more deeply. This—this right here is what made Einstein the most famous scientist in the world. It wasn't his successes, his quotable quips, or his zany hair—that was all icing on the cake. Einstein lived out in earnest what all scientists claim to do but secretly do not—he humbly questioned his own assumptions, examined his own understanding, and constantly strived for improvement. It was this level of humility that led to his successes and ultimately ushered in the era of entanglement.

Einstein's intellectual cravings were for *unification* of his theory of relativity with quantum physics. To achieve this, he had to get to the core assumptions and consequences of the theory. He found that quantum things, like atoms, can be correlated. How? Well, in practice, it's complicated and requires a few hundred million dollars' worth of lab equipment, but let's paint a caricature of it. Take two atoms and rub them together. Now, seal them up in boxes and separate them—send one to Toronto, Canada, and the other to Sydney, Australia. Have some property like the position in the box be measured. Lo and behold, they have the same position! They are correlated. So by looking at the atom in Sydney, you can infer exactly the position of the atom in Toronto. Cool, cool. No big deal—that's how correlations work. But you could have also looked at the momentum of the atom in Sydney and inferred the momentum of the atom in Toronto. That the atoms are correlated in all their measurable properties is what physicists now call *entanglement*.

So entanglement is like super-duper correlations or something, right? That's neat, but what's the big deal? If you have been paying attention (in particular to chapter 3), this should bother you at least a little bit. The idea that we can infer either the position or the momentum of the atom in Toronto perfectly seems to run afoul of the uncertainty principle, which said—in no uncertain terms—that we can't do precisely that. To keep the uncertainty principle intact, the only alternative is that the atom in Sydney somehow influences the atom in Toronto, forcing it to be the same. Einstein called this "spooky action at a distance."[ii] He rejected this idea because it violates the principle that nothing can travel faster than the speed of light. To Einstein, then, this all suggested that quantum physics was not a complete theory.

After a *New York Times* headline in 1935 read "Einstein Attacks Quantum Theory," attention to this feature of quantum physics was prompt. Schrödinger was the first to name it. He called it *Verschränkung* (pronounced ver-shraw-rank-ah-fuck-it-who-cares) and translated it himself to English as "entanglement." Of it, he said entanglement is "the characteristic trait of quantum mechanics, the one that enforces its entire departure from classical lines of thought."[iii] Then he got so pissed off about it that he fucked off to become an Irish biologist. True story.

After some public arguments between Einstein and Bohr, most physicists forgot about entanglement, and I'm pretty sure the public never really cared in the first place.

Local News First

While quantum entanglement has to do with correlations, it is usually talked about in the context of *reality*. Reality means something different to physicists than it does to the average person. To the average person, reality—or what is *real*—is what they can directly perceive with their senses. Tables, chairs, the smell of coffee, the hum of an air conditioner, the clacking of keys on a keyboard...am I just looking around the room I'm sitting in, naming things I see? Yes, yes, I am.

But as an average person with everyday problems like the car being overdue for service, I don't think about what is *really* real. The table, my physics brain tells me, is mostly empty space, occasionally dotted with atoms, which are also mostly empty space dotted with the occasional fundamental particle. But are those particles *real*? I can only really know them as parts of mathematical equations in some scientific theories. This starts to edge a bit too close to philosophy, which is dangerous. Practitioners and students of technical disciplines are usually warned against philosophical questions because they don't directly help solve practical problems. But the real reason—which no one is ever told—is that physicists don't realize how much they don't know about philosophy. So when they start lecturing about reality, actual philosophers cringe like they are judges on a talent show forced to watch a metaphorical train wreck conducted by an arrogant assclown. So to avoid being an assclown, I will not preach here on what physics reveals about the ultimate conceptions of reality.

What physicists can say with confidence and accuracy is what reality is like in our *models* and *theories* of true reality— which we don't have absolute access to. The model we are most comfortable with is Einstein's relativity, where reality is like a theater play. The stage is called *space-time*, the $(3 + 1 = 4)$ four dimensions of space and time, and the actors are the matter and energy that act on this stage. The most important thing in this theater is that causes and effects require contact between things at some point in space-time. Physicists call this *locality*.

Suppose some character in the play chokes and dies from a drink because it was poisoned. Let's dial back time. She must have taken the drink from somewhere. Wherever that was, the culprit must have also been the one who put the poison in the drink. Mystery solved. So although the villain didn't drip poison directly in her mouth, we can trace back a chain of events to a single point in space and time that implicates him—that bastard! I'm so invested in this story. Of course, the story might contain some magic and ghosts or whatever, but no one really believes in that shit. Oh, they do? Well, fuck.

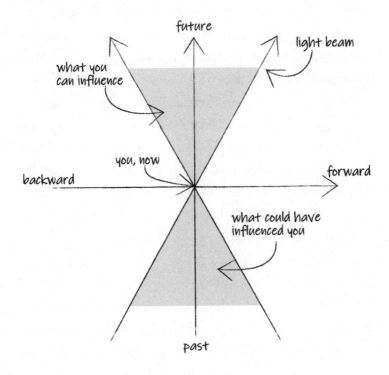

In the nonmagical world Einstein gave us, reality is a collection of events strewn across space-time. The events that can influence other events are the ones that could have been located nearby at some point in time. That sounds like it would be complicated to find out, but it isn't. Take, for example, you. Yeah, *you*, you beautiful beast. What events could you have influence over? Well, you could step in any direction. So from where you are, imagine a circle around you with a one-step radius. Within one step, you can influence anything in there. But there are things that can travel faster than your leg. The fastest influence—or information—can travel is the speed of light. So really, in less than a tenth of a second, your area of influence is the entire globe! By sending information with light signals (which really is how you do it using the internet), you can influence events anywhere around the globe in a fraction of a second. In fact, if you happen to be a TikTok influencer, you probably picked up the wrong book. But also, you'd really be able to influence events this quickly. But the robots living on Mars, for example, cannot be influenced in less than three minutes—the time it takes light to travel from Earth to Mars when they are closest. So if Elon Musk tweets "i coming to mars," and a few seconds later, the *Curiosity* rover cheers (or probably shits its metal pants), that has to be a coincidence, right?

Well, not exactly. Musk could have paid someone at NASA many years ago to program the rover to cheer at that exact moment. So not only do we need to consider the sphere of influence of one event into the future, we need to expand that sphere into the past. If the spheres of influence of the two events intersect at any point in the past, there could have been something at that point that caused both of them. For entangled quantum shit, this is not possible. Events can be perfectly correlated with no possible common past cause. In other words, they don't respect Einstein's locality. So physicists say that entanglement is *nonlocal*.

All this really says is that quantum physics demonstrates that our classical theories are not perfect, but we should have already known that. We need to revise our concept of what cause and effect are at the extremely subtle scales where entanglement acts. It's nothing more than a slight annoyance, really. It's like buying a pack of bagels and finding out when you get home they aren't presliced. Wow. Super inconvenient. This is not the reality I wanted. But what am I going to do, starve to death? No. I'm going to skip the bagel and just eat the fucking peanut butter out of the jar. What does this have to do with quantum physics? Nothing. The lesson here is don't sell bagels that aren't sliced.

The lesson for quantum mumbo jumbo is to not trust scientists when they make sweeping claims about the entire universe because a few physicists weren't comfortable with a toy model of it. Yes, entanglement challenges older theories that suggest reality is a nicely linked chain of causes and effects—but who the fuck cares? Those theories are just as wrong as every other pretty mathematical picture we create of the world. Get over it.

Fast-Forward

Today, an undergraduate physics student cannot get away *without* learning how quantum entanglement works. They may even perform experiments in their laboratory classes demonstrating its effects. How did we go from mocking Einstein's lamentations about entanglement to having children press buttons to create it? The simple answer is technological progress. Even things like the idea of isolating a single atom were thought to be impossible to early quantum physicists. Basically, the things Einstein was worried about were considered to be forever moot and philosophical questions at best by most of the scientific community.

There was a transitionary period in the 1970s and 1980s when single particles of light could be created and individual atoms could be trapped. The fundamental questions were still considered curiosities by most physicists though. For example, one of the most influential people in quantum entanglement theory was John Stewart Bell, who had to do his research on the topic while on leave from his "real job" of high-energy particle physics calculations. Asking fundamental questions in physics is still considered by some to be a career death sentence—do it on your own time, beatnik! There's a saying that was commonplace in physics back then that goes, "shut up and calculate." And calculate they did.

The 1990s and 2000s were the golden age of entanglement theory. Not much was learned that would have gotten Einstein excited; we're really no closer to his ultimate goal of merging quantum physics with relativity. But fuck did we make a lot of technological progress. I mean, I can beckon a complete stranger to bring me food by smashing my fingers on a portable magic screen that probably knows I want kebabs even before I do. I guess laboratory science has made some progress too, but more importantly, who wants kebabs?

The funny thing about quantum physics is that everything we understand about it could have been discovered the day Schrödinger wrote down his equation going on one hundred years ago. All the mathematics is in there. But it takes seeing it, touching it, and building it to really appreciate those details. Nowadays, generating entangled atoms and photons is routine. All those early quantum physicists would shit in their pleated trousers to see what we could do today.

But that all being said, entanglement is no closer to our everyday reality. It is fickle and fragile and ultimately still microscopic. You cannot wield it or sell it in a can, and even if you could, what bullshit could you possibly be selling with something that is random but subtly correlated with something else? Nothing. Unless it is going to slice this bagel, I'm not interested.

Time to summarize.

In classical physics, correlated events cause one another or can be traced back to a common cause that could have determined the outcome. In quantum physics, correlated events have no common cause that can determine their outcome.

The Greatest Joke Ever Retold...Again

Heisenberg is driving... Huh? Oh, you've heard this one? Well, I bet you haven't heard *this* version. Heisenberg is driving with Schrödinger *and Einstein* when they get pulled over. The officer asks Heisenberg, "Do you know how fast you were going?"

Heisenberg replies, "No, but I know exactly where I am."

The officer looks confused and says, "You were going 150 kilometers per hour!"

Heisenberg throws his arms up and cries, "Great! Now I'm lost!"

The officer looks over the car and asks Schrödinger if they have anything in the trunk. "A cat," Schrödinger replies.

The officer opens the trunk and yells, "This cat is dead!"

Schrödinger angrily replies, "Well, it is now!"

The officer shouts, "All right, which one of you do I need to bring to the station?"

Einstein looks at Heisenberg and Schrödinger and asks, "Roll dice for it?"

Hahahaha. It gets funnier every time. But science is no joke—stop laughing. So let's go back to where we started, with *love*. Pfft.

The Fifth Dimension Is Love

Were you aware that in addition to the physical dimensions in which physics plays out, there are *spiritual* dimensions? I wasn't. That is, I wasn't until I visited the internet.

So we all know there are three spatial dimensions, right? Most people learn that the fourth dimension is *time*, which combines with the other three to make *space-time*. But that is just what scientists say. The *internet* tells me that—no—the fourth dimension is that of dreams. Moreover, there are many more spiritual dimensions that you can access through meditation, hypnosis, and credit card transactions. You want to keep your body in the fifth dimension or higher, apparently. Why? Because the fifth dimension is love. It's true—I even saw it in a highly acclaimed blockbuster movie called *Interstellar*. The internet *and* Hollywood can't both be lying, can they?

Time travel is as immensely complicated as it is impossible. We want to believe it is possible, but the concept is rife with contradiction and paradox. This confusion leaves Hollywood scriptwriters a lot of leeway for crafting plausible-sounding bull-shit. *Interstellar* does it in a unique way. Basically, the movie is about a dad and his daughter, whose room is haunted by a ghost that pushes books off her bookshelf. The dad has to go to space. There are black holes and shit. He goes into a black hole, which is really a fucking time machine, and ends up behind his daugh-ter's bookshelf where he—spoiler alert—is the one who pushes the books from the shelf! Whoa, twist! But how does he get there? The *fifth* dimension, of course. But why end up in his daughter's room? Love, of course.

Truth be told, there is a lot of cool science in the movie, but the mechanics of the plot twist aren't the most accurate. In fact, they just leave the audience with more questions. Science journalists, and even scientists themselves, took to the blogo-sphere and social media to speculate on what love has to do with the fifth dimension. The consensus? You guessed it—quantum entanglement!

Every piece of fiction that wants to use science to explain why some plot device must employ instant effects across large distances in space and time invokes quantum entanglement in one way or another. But where does the misconception that quantum entanglement works this way come from? You'll be surprised to find out that, this time, it isn't hucksters of quantum bullshit—it's actual fucking scientists!

Quantum Tickling

Separate two particles, even at opposite ends of the universe, and your choices affecting one will instantly affect the other. That's actually not true. But you will hear this sort of shit every time entanglement is mentioned. You don't even need me to summarize the bullshit I've been rotting my brain on for the past several months while researching this trash. Other thinking, breathing, grown-ass, educated adults are saying the same thing.

Here are the receipts. Let's start with *Time* magazine. I mean, surely a magazine with the name *Time* can't get time fucked up too much, but let's see:

> Once that quantum union had taken place, the rest of the entanglement process played out—with observations or measurements of the spin rate of one electron instantly affecting the other one. And in this case, instantly means exactly what it is supposed to mean.[iv]

Is it supposed to mean obeying the laws of physics? Because "instantly" is not in the lexicon of physics. Speaking of the *laws of physics*, let's see what an actual scientific periodical has to say, and which could be better than a journal called *Science* itself. *Science*, by the way, is one of the most reputable academic journals. Reporting on a recent peer-reviewed publication, it had this to say:

> One of the strangest aspects of quantum physics is entanglement: If you observe a particle in one place, another particle—even one light-year away—will instantly change

its properties, as if the two are connected by a mysterious communication channel.[v]

Nope. Fuck off with your mystery. This isn't Scooby-Doo. How about a reputable publicly funded news organization like the Australian Broadcasting Corporation? Surely they can't be misleading the public in their reporting of recent science news, right? Let's check:

Dutch scientists say they have proved the effect is real, and that simply observing one particle can instantly change another far-away object.[vi]

Oi, mate! Ya got a kangaroo loose in the top paddock? Facken ell! (That's Australian for *fucking hell*, by the way.) Just because a Dutch scientist says it doesn't mean it's true—unless it is beer science (they would never lie about beer science). Let's check in with "all the news that's fit to print." *New York Times*, what say you?

In a landmark study, scientists at Delft University of Technology in the Netherlands reported that they had conducted an experiment that they say proved one of the most fundamental claims of quantum theory—that objects separated by great distance can instantaneously affect each other's behavior.[vii]

The Dutch again! Sigh. Well, I guess I can take solace in the fact that the media hasn't turned misinformation about quantum entanglement into more political tinder for the dumpster fire we are currently burning our minds and planet with. One last attempt. Maybe a popular science magazine chartered to supply accurate and insightful scientific facts to its audience will fare better. *Scientific American,* which I'll admit I subscribe to, had this to say:

> Entanglement is a consequence of the strange probabilistic rules of quantum mechanics and seems to permit an eerie instantaneous connection over long distances that defies the laws of our macroscopic world (hence Einstein's "spooky" remark).[viii]

Oh, fuck off! By now, this should have made your jaw drop like you drank too many radium shots. These aren't just harmless, whimsical summaries of real scientific discoveries. It might be if it were easy to find and digest sensible overviews of the research, but those don't exist. So here we are with the universe's greatest compendium of knowledge at our fingertips, and all we can find out about entanglement is lazy claptrap.

So now imagine if you were serious about trying to make sense of the human emotional state referred to as love. It is quite a complicated and mysterious phenomenon. People "in love" do speak of a "connection" where no physical mechanism can be found. Perhaps it is just magic that you can't explain with physics. Or maybe you can? Let's look into the science literature... Oh hey, what is this entanglement thing? This guy here says it is a magical connection between two things, and he has a Nobel Prize! Bingo! Love is entanglement. This is why we can't have nice things.

I Don't Believe in Science

The Canadian Broadcasting Corporation headline reads, "Neuro Connect claims its $80 clips use quantum entanglement to boost your wellness."[ix] Sigh. This isn't even bullshit—it's blatantly obvious bullshit. By now, you should be thinking, *We know. Why are you telling us this, Chris?* This case is quite illuminating because you can watch how otherwise intelligent people (if you can call venture capitalists intelligent) can be dazzled by science-sounding nonsense. On the television show *Dragons' Den* (the Canadian equivalent of *Shark Tank*), chiropractor Mark Metus convinced a group of investors to give him $100,000 for a fucking paper clip.[9] He says it's made with quantum entanglement, and everyone jumps up to buy in. One investor said, "It's called science. I love the product, and I strongly believe in science." These are investors with hundreds of millions of dollars of net worth—so presumably not complete dumbasses. But I'm sure every person who read this news story thought that *they* could spot the fraud where the expert business people could not. Ha! Maybe if they read *this* book, but otherwise, don't kid yourselves.

Science is not something to "believe in." There is actually a different name for this—it's called *scientism*. Scientism helps no one. In fact, I'd say that if you categorize yourself by what you *believe*, then you can be easily fooled by someone exploiting those beliefs. On the other hand, to be a true scientist doesn't require you to *believe in* any particular thing. In fact, it requires you to constantly evaluate the beliefs you might have in light

---

9 This is the only YouTube video I will actually recommend in this book since it is from a legit source. Otherwise, avoid social media like the fucking plague it is. "Scam or Science? How Not to Get Fooled (Marketplace)," CBC News, February 2, 2018, YouTube video, 22:25, https://www.youtube.com/watch?v=P-Kl0XkZuCw.

of new arguments and evidence. In effect, as a scientist, your beliefs are always temporary.

So remember, next time you question some science jargon–laden bullshit and receive the retort, "What, don't you believe in science?" don't be rude. Simply reply with "no." And then kick them in the shins and run away. You are going to have quite the reputation for this.

What, you're still here? I suppose you want me to say something positive about quantum entanglement. Fine. Remember the chiropractor with the clippy thing? Well, there is actually a kernel of truth to what the quack said. If we are generous—which we aren't—we would say his claim about "reconfiguring the atoms" in the paper clip is not far off from what quantum physicists and engineers actually do. Except we don't do it with paper clips.

When referring to materials, the term *natural* means things *not* made by humans. Though we can create materials that are found in nature, we can also make things completely unknown to nature. This was done for thousands of years by batshit-crazy alchemists. Now material scientists are truly gods, able to create things one atom at a time, like building reality from its basic building blocks. And thanks to quantum entanglement, we can make some pretty weird shit.

Superconductors have been around for many decades. These are quantum materials that conduct electricity with zero resistance. They have to be really cold before the effects of superposition and entanglement are apparent, which is why they are cooled with liquid nitrogen or liquid helium. Superconductors are widely used for making strong magnetic fields. If you have ever seen an MRI (magnetic resonance imagining) machine, then you have seen a liquid-helium-cooled superconductor. If you have ridden on a maglev (magnetic levitation) train, you've sat on liquid-nitrogen-cooled superconductors. Or perhaps you have seen that dead frog being levitated above a superconductor on the internet. I hope it was dead anyway.

There are also super*fluids*, which—analogous to super-conductors—act as fluids but with zero viscosity. If you create a vortex in a superfluid, it will spin forever. Superfluids will also creep up and over the walls of the container they are in. That's about as close as you can get to real magic. What, don't you believe in science? Just because you can't use quantum entanglement to help your golf swing doesn't mean it's useless and boring.

6

Infinitely many goddamn worlds

What is real? Do we have free will? What is the answer to the ultimate question of life, the universe, and everything? No, it's not forty-two.

In Douglas Adams's *The Hitchhiker's Guide to the Galaxy*, a society of advanced intelligence builds a supercomputer specifically to answer this question. It takes the computer over seven million years to compute the answer, which the computer reports is forty-two. And although that's all most people know of the story, the computer goes on to illustrate a very deep and important point—the answer is meaningless because the beings couldn't even understand the fucking question they were asking.

This is how I feel a lot of the time when reading about what quantum physics is purported to teach us about reality. It's not that the answers are wrong so much as the questions being asked don't even make sense. There is a phrase for the nonsense that can't be confirmed or refuted—it's called "not even wrong." We've simply been calling it bullshit. The answer to all the questions raised or implied above is *we have no fucking clue*. Like, we're not even close. These probably aren't even the questions we should be asking. It's not that the answers we have aren't right, it's that they're *not even wrong*.

On the other hand, what else are a bunch of sacks of meat and bacteria supposed to do? If nothing else, it's fun to speculate. But I have to be very fucking clear here. If you read a sentence that begins, "Quantum reality teaches us...," take the rest of that sentence with a generous helping of salt—like a stroke-inducing amount of salt. There is nothing in the equations of quantum physics that says anything about what reality *is*. But like a philandering politician might say, that all depends on what the definition of "is" is.

Reality Bites

For a scientist, the definition of "is" is what is *real*—as in objective reality. Luckily, for our ancient ancestors, what was real was what was presented to our senses—or what will be presented to our senses when that lion jumps out of the fucking bushes! But then humans moved into cities and had the comfort (or curse) of free time. When we stopped having to worry about basic needs like survival, we started philosophizing, and it's been all downhill from there. Fast-forward a hundred thousand years from when we emerged on the savanna, and today, "real" means whatever happens outside the internet #IRL. How did we get to our boring-ass dystopia?

Nowadays, we categorize thoughts on reality into *realism* and—unimaginatively—*antirealism*. Realism is the idea that things exist independently of humans, our beliefs, our perceptions, and so on. There is a real world out there. It existed before us and it will exist long after we're all fucking dead—which doesn't look to be far off at this point. Realism comes in many flavors, which we will get to soon. Antirealism, on the other hand, says nope to all that shit. It also comes in many flavors, but generally it says that reality is *not* independent of who's asking about it or that it really doesn't matter since the question is irrelevant—like asking what a square tastes like. Speaking of taste, what about these "flavors" of philosophy? That sounds fun—like they all have their own version of Ben & Jerry's ice cream. Come try Icecreamential Nihilism, an empty tub that can be any flavor you want it to be! Unfortunately, it's not that fun. Instead of ice cream flavors, they are more like colors of paint you'd like to watch dry. Let's get started.

Realists agree that something exists independently of our perceptions of it, but they don't agree on what that something is. Interestingly, the first notion of realism is also one of the most extreme. Plato, of ancient Greek toga party fame, considered the world we see as an imperfect representation of "forms." An example form is a circle. There are no perfect circles we can see, touch, taste, smell, or feel. For Plato, the circle occupies the *real* world along with other abstract stuff. All these poser circles are just imitators. You can tell it was good philosophy because—like good art—it allowed people to argue about it for thousands of years.

But if we have moved on from ancient philosophy, why do I bring it up? Well, for one, there are only two chapters left after this, and I was contracted to write fifty thousand words. But more importantly, I will need to warn you away from a particularly extreme tendency to see deep meaning in mathematical symbols in equations, which is a sort of naive kind of Platonic realism.

In the intervening centuries, something really important happened that no one today gives a shit about anymore—science.

It Works, Bitches!

What is red? Is red *real* and things that *appear* red are only imitations of a pure idealization of *redness*? Or does the *idea* of redness emerge from the reality of red things? This is the kind of shit people used to argue about. But before you chuckle at how juvenile this sounds, google "the dress" and enjoy the inanity of watching your fellow modern-day compatriots argue about pointless horseshit.

Today I can tell you that a red ball is red because the species and arrangement of the molecules of which it is made have a quantum energy level structure scattering light that matches the frequency we perceive as red. Science. It works, bitches. By the way, that wasn't supposed to be some weird flex where I show you up with my science knowledge—although I am feeling pretty self-righteous about it.

The point was to demonstrate the seemingly final authority scientific explanation has. It is powerful now, and it was powerful hundreds of years ago too. With empirical science came the idea that investigation allows us to discover the true nature of the world. Successful scientific discoveries are elevated to "laws," after all.

Natural laws are the most extreme consequence of scientific realism, which states not only that reality exists independently of us, it also has a set of mechanical rules that it follows. In this view, the universe is a large machine that chugs away, never straying from its internal perfection. As scientists, our goal ought to be in understanding the mechanisms that govern the great machine. It's a powerful idea, no doubt, and it guided the course of physics for hundreds of years. But let's throw a fucking wrench into this beast.

Quantum Kludges

We've already learned a lot about quantum physics so far. And when I say *we*, I really mean that I already knew a lot, and I told you about it over the last five chapters. But five chapters and 32,515 words are a lot to take in. Yeah, I know, I'm really conscious of the word count. But you can only say fuck so many times. Anyway, as I've told you, quantum physics was born at the turn of the twentieth century. But it wasn't as *clean* as even I have presented it so far.

You see, those people we call *quantum* physicists—like Planck, Bohr, Heisenberg, Schrödinger, and so on—were not quantum physicists at all. They were classical physicists! They were classical physicists struggling to explain the new observations from experiments exploring the microscopic world. Of course, every attempted explanation was phrased in the only science they knew—classical physics.

For every new crazy idea that we now remember and have given Nobel Prizes out for, there were probably dozens, if not hundreds, that simply did not work. How can you correctly state a fact in a language not yet invented? Either you can't or you need to awkwardly do it while inventing the new language. The language of quantum physics is a new kind of mathematics. But I know what you are thinking. *Do not write even one more fucking equation, Chris. I swear to fucking god I will burn this goddamn book!* Don't worry though; equations are too much of a pain to convince my editor to include, and as you can tell, I'm getting pretty lazy at this point.

Now that I have you thinking how much you hate math, a question that is probably sitting in the back of your mind is *why is math so important in quantum physics anyway?* Well, truth be told, we kind of wish it wasn't. Let me explain.

Remember Newton? He was that guy who avoided the plague and invented gravity. Newton was all like, *yo, check this shit out. Gravity is a thing, right? Well, I can write some math down to make it way more convenient. Who wants in?* And everyone was like, *okay.* So the math in classical physics was built on top of the existing mechanical conception of reality. Apples fall from trees. No shit, but *where* is the apple? Newton showed us the way through with mathematics by starting with the statement "let the variable x be the position of the apple."

Now I hope this doesn't trigger any trauma from the dreaded word problems that everyone hated in elementary school math. The point is only to show that no one is going to look at an equation with x in it and not understand what it means—it's the position of the fucking apple! We can all point to where the apple is—there is no controversy. The math in classical physics is in direct correspondence with our everyday conceptions of the world.

The trouble with quantum physics is that math was invented to match observations that no one could understand using classical physics, let alone their intuition.

English Words for a Quantum World

Do you remember when I told you about quantum jumping? It's okay if you don't—I didn't either and had to go back and reread it. (Chapter 1, by the way.) The TL;DR is that atoms have discrete energy levels for an electron to occupy. To change energy, the electron gives or receives exactly one quantum to "jump" between energy levels. Fine. That all sounds plausible when it is told to you by a quantum physicist...or does it? It "jumps?" How? How long does this jump take, and where is the electron during the jump?

These questions exemplify the trouble with *talking about* some reality behind the observations and math of quantum physics. Niels Bohr's opinion was that you can't even ask these kinds of questions—there is no *quantum world* behind what we observe in the experiments. This *antirealism* view was summed up as the advice—or demand—to *shut up and calculate*. If Bohr was right, then using words like *particle* and *jump* is a bad idea because they conjure up an image of something, well, jumping. But what else can we do? We have to *say* something. You might think that perhaps the math that I keep raving about can simply speak for itself. But no—that apparently only makes things worse.

The *new* quantum theories ushered in by Heisenberg's matrix mechanics and Schrödinger's wave mechanics were also desperate attempts at fitting math to observations. The trouble was, they worked *too* well. After de Broglie suggested that matter—like particles of light—could have wave-like properties, Schrödinger thought, *hey, if there is going to be some wave equation, maybe it could look like this.* Then he wrote down his equation and it worked. Shit. Perhaps it was too easy.

The symbol Schrödinger used in his equation is the same

one we use today, the Greek letter psi, ψ. But unlike the phrase "let the variable x be the position of the apple," there was no phrase "let the variable ψ be the [insert something conceptually simple]." The Schrödinger equation can be used to make predictions about the position where an electron will be measured, but ψ doesn't represent the position of an electron. The Schrödinger equation can be used to make predictions about the momentum or energy of a measured electron, but ψ doesn't represent the momentum or energy of an electron. And so on it goes. We can use Schrödinger's equation to make predictions—the most precise predictions of any scientific theory ever, in fact—but it doesn't tell us anything about the meaning of ψ.

Today we understand ψ as the *state* of a physical system. Remember that the state is a summary of everything that can be known about a system. To know the state of something means to be able to make accurate predictions about what will happen to it. That is what ψ is. It's been called many names, including the psi function, the wave function, and also just the quantum state. As a quantum physicist, I use quantum states every day. I use them in calculations to make predictions about the outcomes of experiments and nothing more. They aren't *real*; I can't use them to detoxify my auras any more than I can use them to prop up my wobbly desk. The object ψ I insert into equations to help solve a problem, and then I throw it away.

Some have demanded more of ψ. Not only should it allow us to predict the outcome of every measurement we could make on the system—the *what* and the *how*—but it also ought to tell us *why*. To many quantum physicists, the answer to *why* must be an explanation regarding some deeper reality. If it is not, they'll just keep asking. Quantum physicists are like annoying little three-year-olds who are never satisfied with "just because, okay. Now eat your fucking broccoli and go to bed."

But...Why?

What does ψ teach us about reality? Answers to this question are called *interpretations of quantum theory*, which are part of a research field that goes by the name *foundations of quantum physics*. Like the foundations of a building, foundations of a theory are important—especially if the rest of the building was built by a bunch of lazy grifters under contract from rent-seeking assholes. What I'd like to say about the analogy to the property market is probably too dark even for this book, so I'm going to cut it loose there. Anyway, the goal of an *interpretation* of quantum physics is to answer all those unanswered questions that people have been asking for the last hundred years. This would be much easier if those fucking physicists could agree.

Well, to be fair, they do agree on the uncontroversial facts. That is, no one disagrees with the results of quantum physics experiments. Everyone agrees with the textbook mathematics that is used to make predictions. That is, no one disagrees with the presence of ψ and the Schrödinger equation. What is ψ *really*? That is the question.

Some say there are as many interpretations of quantum physics as there are quantum physicists. But as far as a hopeful eventual consensus goes, there are only a few contenders. There are the boring-ass ones no one hears about and the crazy ones that get headlines in the popular science magazines. You can probably guess which I prefer.

Interpretations of Quantum Physics
You'll Never See in a Movie

The oldest interpretation of quantum physics is still the default. It's the one in all university textbooks on quantum physics. It's not difficult to describe because it says nothing at all. This is the *don't ask* interpretation. It's also been summarized by the phrase I've mentioned quite a few times already—*shut up and calculate.* Use the math, solve the problems, don't complain. We could have just as well called it the *fuck off* interpretation of quantum physics.

Ding! Ding! Ding! You guessed it. This one is my favorite. Johnny, tell them what they've won! Johnny? No? Okay, moving on.

Closely related to this fuck off interpretation is the so-called *Copenhagen interpretation.* It's called by the name Copenhagen because that's where Niels Bohr worked when he hosted Heisenberg and other scientists who made the initial developments of the theory. Their philosophy is summarized as *instrumentalism,* which is a kind of antirealism.

Instrumentalism has nothing to do with musical instruments—that would be too cool for philosophy. Instrumentalists believe that ideas, including theories of physics, are *instruments,* tools to achieve some purpose. They are usually silent about the purpose, which is weird because the purpose of *musical* instruments is very clear—use them to get famous enough to afford a constant supply of meth, which brings me back to Heisenberg.

You can imagine the kinds of questions scientists asked after Heisenberg introduced his uncertainty principle. If position and momentum cannot be known simultaneously, does the particle *really* have a position? The answer for instrumentalists is more than *don't* ask—it's you *can't* ask that question. For the instrumentalists, the theory is about what is observed during measurements and says nothing more. You can't ask if the particle really has a position because the theory has nothing to say about it. To instrumentalists, asking for an interpretation of quantum physics is like being given a hammer and nail, being shown how to use them, and then asking, "But what is the true essence of the hammer?" Who cares? Just hit the fucking nail.

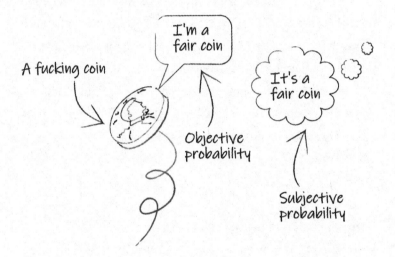

A new interpretation on the hip new quantum foundations scene is called *QBism*, which stands for quantum Bayesianism.[10] It's named in analogy with Bayesian probability, an interpretation of conventional probabilities as *subjective*. A good way to find out if you are Bayesian is to ask yourself whether a "fair" coin is a statement referring to something physically true about the coin or a statement about your own personal expectations

---

10 Pronounce it *cubism*, like that art thing you might find if you don't have kids and can enjoy culture.

for what will happen when it is flipped. If you lean toward the latter, you're a fucking Bayesian—good on ya! The QBists, as they call themselves, believe that ψ is subjective information just like probabilities are subjective information about the expectations of decision-makers. It's a good interpretation and a close second on my list after the fuck off interpretation.

Einstein favored the *ensemble interpretation*, which states that ψ is the quantum state of ensembles or collections of things instead of individual things. He famously claimed that more needed to be added to quantum physics to make it "complete." One such "completed" interpretation is the *Bohmian interpretation*, named for David Bohm. I'd love to tell you about it, but I'm worried that all these names for esoteric philosophical positions that few care about may be a bit much to take in while you're sitting on the toilet. That is what you're doing while reading my book, isn't it?

Let's just cut to the chase then. The one you've all been waiting for...

Parallel Fucking Universes

Hugh Everett...sorry, Hugh Everett III, was an American physicist who missed out by one generation from being Hugh Jr., which is the coolest suffix to have. "Sr." is just lame because anyone can give that to themselves by acquiring and naming a kid after them. Some sequels are better than the original, but the threequel always sucks—*Home Alone 3*, anyone? True to form, Everett III led a relatively disappointing life.

His PhD thesis was devoted to a new interpretation of quantum physics, but no one took him seriously. On one occasion, when visiting Bohr to present his ideas, another physicist said he was "indescribably stupid and could not understand the simplest things in quantum mechanics."[i] Ouch. That doesn't seem fair, but it does foreshadow the controversy that continues to surround Everett's theory. It goes by several names, the most popular of which is the *many-worlds interpretation*. However, champions of the interpretation more often refer to themselves as Everettians. I guess all cults need a leader after all.

Cosmologists are the physicists who study the universe *as a whole*, including its entire history and potential future. Cosmologists try to reduce all that complexity to elegant mathematical theories—they beautify theories of outer space. They are not to be confused with cosmetologists, who beautify one's outer *face*. Don't make the mistake of buying someone a gift card to see a cosmologist. They will not thank you for it.

The most famous part of cosmology is the Big Bang theory, which states that the universe was born fourteen billion years ago when—out of nothing—space started to rapidly expand. Clearly, there were no quantum physicists around at the beginning of the universe, privately assigning quantum states to things in their labs. Presumably, there was one unique quantum state ψ describing everything when the universe's stopwatch showed 0. This is sometimes called the *universal wave function*. Once the stopwatch started, the universe and its quantum state started evolving according to the quantum laws set by the Schrödinger equation. Everything that has happened and will happen are all encoded in ψ and how it changes over time from 0 at the Big Bang.

Fast-forward to now, and the stopwatch shows fourteen billion years. There appear to be people observing the world and finding it strange that things act wavy until measured, at which point they act like particles. But these people are mistaken. There is no such thing as "measurement." There is only ψ—and it contains a massive superposition of everything that could have or can happen. Admittedly, it's hard to think about what this actually means. But there is a cheap way to imagine it—parallel worlds.

Whenever something happens that looks to an observer like a measurement, the universe "splits." It is now as if there were two universes. In one universe, the observer sees that Schrödinger's cat is alive, and in another *parallel* universe, the observer sees that the cat is dead. I need to remind you that—even in the wild views of Everettians—there are not really two universes. There is only one wave function and hence one universe. A *split* is a quantum superposition, and these can be undone. Quantum physicists can routinely demonstrate this with atoms and photons. In the language of Everettians, the universe splits and is then recombined, but at no point do the "universes" interact with each other.

The difference between placing an atom and photon in superposition and placing a cat and observer in superposition is the amount of information involved. *In principle*, if Schrödinger's gruesome experiment could be carried out, it could also be undone. However, in practice, this will never happen as the information that differentiates each part of the superposition would be impossible to collect and process. Think of it as throwing an encyclopedia into a fire and trying to collect all the smoke and ashes to retrieve the information "lost." It's possible *in principle* but will never happen in practice.

So there is one universe, but it contains a superposition of a live cat and happy observer as well as a dead cat and some explaining to do. Each cat/observer combo is in its own universe in the sense that the superposition will never be recombined. It is as if there are two universes, each with its own observer who is identical in every way, except one has a mess to clean up.

If you've followed up to now, then the next step is to understand that splits are happening *all the time*. And if branches of the wave function had fourteen billion years to independently evolve, they could be quite different indeed. These vastly different universes—perhaps even with different laws of physics—are sometimes called *the multiverse* by popular scientists who love to talk about it in sultry, deep voices. (The multiverse is a name also used for other cosmological ideas. Stay tuned for my next book, *Fucking Funky Cosmology*.)

All right, that's enough. Can we *prove* that any of this is true? Nope—at least not in the sense of being able to perform an experiment that differentiates one interpretation from another. Remember, they are all ways to make sense of the same theory and hence make the same empirical predictions. Why bother then? Well, there is also that nagging question of *why*. But some physicists think that understanding the "correct" way to interpret quantum physics is the only path to progress in science. Me? All I know for certain is that I could use a beer.

In classical physics, the interpretation of quantities is so obvious and clear that no one talks about it. In quantum physics, the interpretation of quantities is still hotly debated, and there is no scientific consensus or basis for forming concrete views of reality using it.

Quantum Suicide

Well, that sounds morbid. And no, it's not some twisted joke or pun—it's an actual term for a consequence of the Everettian interpretation. This is going to get a bit gloomy though, so I need to pause here for a warning. Think of it as a disclaimer of sorts. I will not be held responsible for any stupid thing you decide to do because you misunderstood quantum physics.

Remember how I told you that Everett led a disappointing life? Well, scientists not taking him seriously led to him leaving academia to work as a defense contractor. Nothing wrong with that, I guess. I mean, defense contractors are always depicted in Hollywood as totally not greedy sadists. Everett led an indulgent life of eating, drinking, and chain-smoking. He died predictably young of a heart attack and, as per his wishes, had his remains thrown out in the trash. A decade later, his daughter committed suicide, requesting her ashes be thrown out in the trash as well so that she might end up in the same parallel universe as her father. Yikes. That's intense.

I'm not sure what Elizabeth (the daughter) believed, but Everett was said to be a believer in *quantum immortality*. Now that *sounds* fucking badass. Unfortunately, in practice, believing oneself to be immortal probably correlates well with other poor mental and physical health choices. Overweight, chain-smoking alcoholic? Who cares—I'm going to live forever in the multiverse!

In the many-worlds interpretation, Schrödinger's cat is not alive *or* dead. It's also not alive *and* dead. There are two universes. In one universe, the cat lives. In the other, the cat dies. Sad. However, the important thing is that *in one universe*, the cat *lives!* Now repeat the experiment over and over. No matter how many times we try to kill this fucking cat, there is always one universe in which it lives—the cat is immortal!

Now consider the same scenario, but with you and a gun— quantum Russian roulette, as it were. The same logic (if you can call it that) applies. There is always at least one universe in which you survive this stupid game. This is the kind of stuff you have to contend with when you let quantum physicists play philosophers. Suffice it to say, the many-worlds interpretation is not popular at parties.

The Most Ill-Timed Joke Ever

Heisenberg is driving with Schrödinger, Einstein, and Everett when they are about to get pulled over. Everett yanks the wheel of the moving car, and it rolls and falls off the cliff. He can be heard yelling as the car falls, "We'll survive somewhere in the multiverse!"

This joke killed over in universe #19498302.

I like to think that in a nearby parallel universe, both quantum physics and Rodney Dangerfield get the respect they deserve. But in the universe we currently inhabit, sadly, neither do. Quantum physics is bastardized to no end in popular culture, all in the service of terrible plotlines and shallow characters.

A major theme in fiction is the idea of *alternate history*. In fact, it is an unwritten rule that every television show that lasts for more than one season has to invoke an alternate history at some point. This has mostly to do with the fact that scriptwriters paint themselves into corners because every episode has to have "the biggest surprise yet!" What is alternate history? Stay tuned next week to find out in a shocking new episode that everyone will be talking about!

In the show *The Flash*, the title character intentionally travels back in time at several points, creating multiple timelines that were intended to form one coherent plot—it didn't work. *Westworld* included several timelines stitched together in a meta plot that you'd have to work out yourself as if it were a homework exercise. But the most egregious example was the popular drama *Lost*, which lasted a modest six seasons. In a plot generously called "complex," we watched the present, the past, the future, and an *alternative* present, a flash-*sideways* that was a parallel world in which the characters in the "main" world were reunited. Lost? Perhaps that was the point all along.

One of the most interesting things about alternate history plots, especially ones that involve time travel, is that the theme developed independently of quantum physics, and even the earliest examples within the science genre of this type of fiction predated Everett by decades. Then the idea of "science consultant" emerged, and everything went to shit.

Academia: Mostly Social Distancing since 400 BC

The idea of consulting a scientist is often approached with the best of intentions. However, there's a reason the phrase "good intentions" is never used in a positive way. The hopelessly naive writer or producer thinks, *hey, let's get a scientist to help make the details of our story about an alien superhero that breaks all known laws of physics scientifically accurate!* The scientist's first reaction is probably revisiting the taste of their lunch. But who can resist the glitz and glamour of Hollywood? Maybe I'll get an autograph from Robert Downey Jr.! Do you think he includes the "Jr." in his signature? That would be so cool. And it only costs my integrity? Sign me up!

If you are a fan of Marvel movies, you can probably pinpoint the moment they started consulting a quantum physicist. Hell, in one of the movies, Paul Rudd ad-libs, "Do you guys just put the word 'quantum' in front of everything?" Yes, yes we do. I happen to know one of the consultants, but I didn't consult them in writing *this* book because I'm too lazy. Also, I've probably offended them a few times already. But here is a totally plausible exchange.

Marvel writers: Is there anything in science that can explain instantly traveling around the universe?

Quantum physicist: No.

Marvel writers: How about time travel?

Quantum physicist: No.

Marvel writers: Well, what kind of science do you do?

Quantum physicist: I am a quantum—

Marvel writers: Ooooo quantum, that sounds sexy. Let's use that word for everything.

Quantum physicist: Sigh. Sure, fine. Hey, is Robert Downey Jr. back there?

As of this writing, the second most successful movie of all time is *Avengers: Endgame*. (The movie rights to *this* book are still available, by the way. Looking at you, Disney.) *Endgame* is all about alternative timelines accessed via the "quantum realm." They try to be really clever about it by—in the actual dialogue— criticizing other time-travel movies, such as *Back to the Future*. How fucking dare they! Michael J. Fox is an international treasure, and if he says he went back in time, then it's true, okay? Anyway, *Endgame*—long story short, the bad guys die, the good guys win, and none of it has anything to do with quantum physics.

The trouble we keep running into is scientists legitimizing pseudoscience. We—sure, I'll lump myself in there—don't do this on purpose of course. We love science and want everyone to be able to appreciate it. But in dealing with especially fiction, we end up misrepresenting not only the facts but also how science works. Hint: it doesn't work by having some lone genius in a lab coat looking for some pattern in events before cracking it in some eureka moment that always ends with some hacked-together device that sends out the "right signal" to "kill the queen" or whatever. That's the plot of every science fiction movie but definitely not how actual science works, which is the exact opposite. Any particular field of science is a collection of thousands of people, each playing a small part in a solution that is built by a series of ever-improving failures.

Scientists can't go from nothing today to saving the world tomorrow. And if you wanted us to find a way to kill off the one species invading and destroying every corner of the planet, I've got bad news for you. Maybe you can hide in another dimension.

Doing Your Own Research

When I started writing this godforsaken chapter, I typed "parallel universes" into a search engine. Instead of giving me results for my query, it adamantly suggested I search for the more commonly asked question "how to contact yourself in a parallel universe?" Oh shit, this ought to be good...

It makes perfect sense to ask this question if you've read or heard anything in popular media about quantum physics. It's also extremely alluring, the idea that there is another "you" out there who made better choices and book purchasing decisions. If everything that can happen does happen, then at least one version of you is very successful, right? So you should be able to just hop over to their parallel universe like it is another lane on the highway. Then maybe they can give you some advice on how not to fuck up your life so much. Alas, this is not how quantum physics works.

But more importantly, why are you on the internet looking for advice in the first place? Bad human. Never do that. Look for experts and ask *them* for advice. You want to know about solving a health problem? Ask a doctor. You want to know how to solve a financial problem? Ask an adviser. You want to know how to solve a legal problem? Ask a lawyer. You wouldn't ask a physicist to solve any of these problems—we have no fucking clue, nor do we care. So why would you believe some dodgy website that their solution is proven by quantum physicists?

You know the old saying "don't believe everything you read"? Well, when it comes to the internet, a better saying is "don't believe *anything* you read, or listen to, or view on YouTube." Meh. It's not as catchy. I'll work on it.

What to Believe When You Can't Believe Anyone

If everyone is wrong, what is right? If there is no objective reality, does truth even have a meaning? And if truth has no meaning, well, we are all witnessing what happens then. The problem with this logic—if you can call it that—is the false dichotomy. There are *degrees* of right and wrong even in science. Here is an example. Is the Earth flat or round? As it turns out, both are wrong. The Earth has a larger diameter at the equator because of its rotation, so it's actually an *ellipsoid*. The Earth is not flat and it is not round; both of these are *wrong*. But one of these is definitely *wronger*. Now you might wonder why it matters which is *closer* to the truth if they are both wrong. The answer lies in how *useful* your assumptions are. You aren't going to get into much trouble assuming the Earth is round. But if you assume it's flat, well, have you met a flat-earther?

If you are going to believe something, make sure it is the *less wrong* thing. But really, you shouldn't believe anything with undying conviction. Even the idea of a flat Earth works pretty well when you are looking at a map of your city. This shows that ideas can be useful without being ready to die for them or argue endlessly on the internet about them, which feels worse than death to witness. This encapsulates a simple philosophy that doesn't require years of studying to understand: everyday pragmatism.

An everyday pragmatist is someone who *acts* on information, not someone who *believes in* it. An everyday pragmatist is "doing the right thing" when the information has been used to serve some useful purpose. They get on with things and don't dwell on arguments that have no practical value. Is the Earth really a bumpy oblate spheroid? Who cares right now? I'm trying to get to the nearest liquor store, and my "false" map will get me there just fine.

The danger with any kind of realist perspective is being forced into states of cognitive dissonance. If you really believe in your heart of hearts that the Earth is flat, then you *have to* believe all sorts of other crazy shit just to keep a consistent story in your head. How do you reconcile the fact that Australia exists and the people there see a different set of stars in the sky? Easy! They are all actors in a government conspiracy!

So if you are a scientific realist, you *need* to see meaning in the mathematics of successful theories. Then you start to say things like quantum physics teaches us that reality is an infinite multiverse of parallel worlds where everything that can happen does happen! For a pragmatist, this is *not even wrong*—it's the point where you just walk away. You don't have to kick them in the shins this time.

Or maybe I'm wrong and there is a better branch of the wave function where I'm an Everettian evangelist and I wrote a quantum vegan cookbook instead of this shit. Yeah, I like that idea. Quantum avocados, anyone?

7

Quantum fucking technomagic

Well, here we are, the penultimate chapter. Penultimate means second to last. Yeah, I'm still subscribed to that word-of-the-day app. The thing about quantum bullshit is that it all happens in parallel with quantum, you know, nonbullshit. Nonbullshit, mmm yeah, that's my jam. Nonbullshit is science and engineering, and quantum physics is responsible for a lot of it. But even when quantum physics grants us the ability to engineer new technology, the bullshitters are still there, clinging on like dried shit to the hair on a cat's ass.

New technology attracts grifters like a child to birthday cake. Mention the idea that quantum technology is a thing, and a few days later, you'll have enthusiasts, evangelists, and even start-up founders singing its praises. And that's not even the worst part. New technology is always used as the go-to metaphor for anything yet to be understood. After the mechanical clock was invented, people began to compare the workings of the body and mind to the cogs of the device. Then the whole universe became a giant clockwork machine. Today, of course, the universe *is* a quantum computer and the brain *is* a quantum computer, and, sigh...the only thing that *isn't* a quantum computer is an actual fucking quantum computer, which doesn't even exist yet!

A quantum computer will eventually exist though. The nineteenth century was the industrial age. The twentieth century was the information age. The twenty-first century will be referred to as the quantum age, or Armageddon, whatever comes first—it's a close race. Trying to engineer technology one atom at a time is cutting-edge stuff. But it's not as complicated as it is made out to be, and you probably don't even need to care about the details. So why are you getting so worked up? Calm the fuck down already.

Quantum Nonbullshit

If there is one thing I want you to take away from this book, it is to recommend it to all your friends. Also, I want you to remember that quantum physics is not mystical and doesn't have magical properties with miraculous benefits. It does, however, grant us immense technological power, and that only comes from a well-understood and mature science. And nerds—a shit ton of nerds.

We already ran into one of the earliest technological applications of quantum physics in chapter 2—the electron microscope. But I want to get to some things that are more familiar to my esteemed readers and cultural connoisseurs—fucking lasers, pew pew pew!

Everyone knows what a laser is, even if they don't know how a laser works. The laser is typical of technology that exists today in that regard. You can easily conjure up an image of a laser in your mind, but do you know who put it there? No, you don't. It was our old pal Albert fucking Einstein again.

The caricature of Einstein's life has become the mold for many physicists. They start out as young rebellious visionaries and then turn into old retired cranks. (Perhaps you can guess which end of the spectrum I am edging toward.) By the time Einstein became famous and moved to America, he was already in his cranky phase, at least when it came to what he spent his time deeply thinking about. But before American celebrity ruined him, he was churning out fresh ideas in physics faster than science skeptics can come up with vaccine alternatives. (There's no such thing as an alternative to vaccines, in case you didn't already know.) And one of the many ideas Einstein dreamed up was the laser in 1917. The first working laser wasn't built until 1960, five years after his death. Einstein never did get to pew pew a laser.

As an interesting aside, I'd like to point out that much of the technology we enjoy today is rooted in ideas that came out of thinking about physics *just for the sake of it*. If the people who invested in scientists like Einstein and many, many others demanded the yearly returns that are demanded of scientists today, we would have none of this technology. Do you really think that Einstein cared or even thought about the future applications of the laser idea, which include zapping the hair out of your ass? Invest in scientists not because good commercial value will come next year but because your grandchildren will have a more efficient way of getting an Instagram-worthy asshole.

I haven't mentioned it yet explicitly, but as I'm sure you

could have guessed by now, the laser is based on quantum physics, requiring an understanding of the statistical nature of photons. Now, thanks to quantum physics, we enjoy zapping hair out of our asses. Thank a quantum physicist for that, but also laser tattoo removal, laser wrinkle removal, laser teeth whitening, laser stage lighting, laser tag, laser cat toys, sharks that shoot lasers from their mouths, laser harpsichords, laser discs, and—holy shit—those are only the completely unnecessary applications consumerism has enabled (and one I totally made up). If I started listing the actual scientific, medical, military, and industrial applications of lasers, I'd need a whole new book. But these are the things that come from an understanding of quantum physics, not miracle cures for your shitty love life. Bullshit plays no part in engineering.

Going Nuclear

On December 2, 1942, at 3:25 p.m. local time in Chicago, someone sitting in a crowded cafeteria farted. Oh, and the world entered the atomic age. One of the events made a sound. In Chicago, scientists created the first *self-sustaining* chain reaction of radioactive uranium. Everyone knows of the Manhattan Project and its ultimate destructive goal, but many positives have also come from nuclear technology.

Today, about 10 percent of the world's electricity is supplied by nuclear power, which is "green" if you want to ignore the waste material we are leaving behind for the next thousand generations. Some of the "waste"—called radioisotopes—is put to good use though. In fact, you have some radioactive waste in your home. Don't worry; you don't need to consult your various

Facebook faux-activist groups. It's supposed to be there, assuming you are invested in not getting yourself killed in the most preventable way possible.

Radioisotopes are used in household smoke detectors. The radioactive substance used is americium—not named for 'murica, pew pew pew (the gun "pew," not the laser "pew"—please appreciate the difference). Americium is named for the *Americas*. (There's more than one, in case you didn't know.) The americium in your home was made by smashing plutonium with neutrons inside a nuclear reactor—no big deal. The emitted radiation is in the form of alpha particles, which have enough energy to knock an electron off an atom. This makes the atom have a positive charge. That is, it becomes an ion, and any radiation capable of doing this is called *ionizing radiation*. The positive ion and the electron fly off in different directions and are detected as a current. But! If a smoke particle wanders in, the current stops flowing because it attracts the electrons.

Basically, the radioisotopes in your smoke detector are constantly ripping air apart. The electronics can see this happening until smoke particles block it. Quantum atom-splitting energy put to work keeping you alive—assuming you've checked your smoke alarm batteries recently. Fuck, get on it. This is quantum physics, not some free ride.

Radioisotopes are also used in the field called nuclear medicine. The use of radiation in medicine goes back to before the quantum physics of atomic decay was understood. The illustrious Marie Skłodowska Curie championed the use of radiation to treat cancer and diagnose other conditions. In late 1915, there was a single radiologist in the world. She built mobile radiation medicine machines for French doctors on the battlefields of World War I. Fucking badass! Nowadays, there are around thirty thousand radiologists in the United States alone. Tell your favorite politician (if you have one), quantum physics creates jobs!

But putting radioactive shit in your body to see what happens is not the only way quantum physics can scan your insides.

Spin Me Right Round

Something I haven't mentioned yet is quantum *spin*. I haven't mentioned it because luckily no one bullshits about it—it's the perfect blend of boring obscurity and importance. Spin is a quantum property of particles like the electrons, protons, and neutrons inside atoms. The metaphor used by physicists is a tiny spinning magnet—mostly because it has the word *spin* in it, but also because *internal* quantum spin does interact with *external* magnets. We can detect spin with magnets if we are precise enough. Spoiler: we totally are.

The Earth is a giant magnet, as you know from using a compass yourself or watching *MacGyver*, but it's a pretty weak magnet. A strong magnet is the kind you see as room-sized medical scanners in hospitals or the unending stream of medical dramas about hospitals. In the screening room, the patient slides into a giant tube, which is actually a big fucking magnet— and I do mean *big fucking magnet*. The strength of the magnet is provided by the enormous coils of superconductors that wrap around it and are cooled with liquid helium to near absolute zero. Why do they need to be so strong? It's to access the quantum spins inside the water molecules of your body. The process is called magnetic resonance imaging, or MRI. It used to be called *nuclear* magnetic resonance imaging, but Midwesterners kept pronouncing it "nucular," and that pissed off the elites too much, so they dropped it. Just kidding. The reality is much more depressing. MRI was invented during the Cold War, and the word *nuclear* was conditioned with air raid sirens to be the scariest word in the English language.

Doctors were afraid patients wouldn't accept treatments that had the word *nuclear* in them. Probably fair. You can take the "nuclear" out of quantum physics, but you can't take the quantum physics out of nuclear MRI! Because of quantum physics, we know exactly what atoms your body is made of, and we can map them out by whispering to their quantum spins using magnetic fields. Science is so fucking rad. We don't deserve it.

But lasers, modern medicine, television, air-conditioning, and a hell of a lot more have nothing on the greatest invention of the twentieth century—the transistor.

Computer Says, "Yes"

As I type this, I'm making use of two different kinds of nanometer-sized devices: alcohol molecules and billions of tiny switches, too small for the eye to see. I know how the alcohol got there, but the story of those tiny switches is so complex that no one human can understand all of it. It's beautiful, really—that we rely now on the network of human intelligence to maintain our standard of living and continue to progress. We are all connected, and not in some wishy-washy New Age bullshit way. The actions of individual people on other sides of the globe cannot be explained by understanding them alone; their actions are part of the collective behavior of billions of people. They are part of the network. Are you? Come, join us. We have beer.

The massive connectedness of today's society is an indirect consequence of quantum physics and the technology it has enabled. Computers obviously play a large part in that story. They. Are. Everywhere. Just within my eyesight sitting here near the window in my kitchen...my car outside is a computer, my phone in my hand is a computer, my earbuds next to me are computers, there are computers in my daughter's programmable light-up shoes that aren't put away, there's a computer in my refrigerator, microwave, and wall clock...hell, there's even a computer in my coffee machine.

Computers can be everywhere because of the miniaturization of electrical switches. In principle, there's nothing your computer can do that a bunch of people with light switches can't do—it'd just take them a really long time. Also, they probably couldn't fit in my coffee machine. Everything about the manufacturing process that creates computer chips has the fingerprints of quantum physics on it. Even the material used— semiconductors—could not be understood without an understanding of quantum energy levels.

All this twentieth-century technology was built on top of quantum effects such as discrete energy levels, uncertainty, and superposition. But what of entanglement? Is that a completely useless effect in quantum physics? Fuck no! Entanglement is ushering in a new wave of twenty-first-century quantum technology. And (please read this under your breath) even the goddamn many-worlds theory has played a part in it.

But Wait! There's More

Your laptop computer or smartphone performs calculations with *binary logic*. Everything it computes is represented as ones and zeros, and these values are stored in various ways depending on their purpose. If the ones and zeros encode data for long-term storage—such as the thousands of pictures of your cat not looking at the camera—then *hard drives* fit the bill. If the ones and zeros are being actively used to solve problems—such as finding the cat's face in a photo so you can face-swap it with your own—then the trusty transistors can rapidly switch on and off as required. And although the technology used today is cutting edge, it is still only used to carry out simple logical steps one by one, albeit really fucking fast. The idea that simple instructions—a *program*—could be sent to a machine to do complicated calculations goes back at least to Ada Lovelace in a time before face-swapping, when a portrait was a painting by an artist who was probably beheaded shortly after for drawing your nose too big; where is Photoshop when you need it? The fact that the "software" predates modern physics is why we call existing computers "classical," while the new ones are called "quantum."

Quantum computers are devices in their infancy circa 2022. And although from an engineering perspective, they are just the next natural step in technological development, they are designed to carry out completely different kinds of instructions than your smartphone can handle. Instead of doing calculations with ones and zeros, a quantum computer uses more complex mathematical objects. These naturally line up with the math used in quantum physics and can be stored and manipulated with things like atoms and photons. Building a computer one atom at a time is not without its challenges, which is why it is taking so fucking long. But if it is so hard, why even bother? I mean, swapping your face with a cat's face is already the pinnacle of computational achievement. What more could you want?

Computer Says, "No"

The word *computer* has a pretty general meaning. In fact, a "computer" used to refer to people who, well, *computed* things. Today, those people are called math nerds, and the term *computer* is reserved for electronic digital technology. The genesis of the theory of modern computing occurred in the mind of Alan Turing, in one of the most tragic tales of heroism ever told. Turing was a driving force in the Allied victory in World War II, which he achieved by designing and building a computer called the Bomb that was able to crack the German Enigma codes. The rest of the story is completely fucked up, but then again, so are all stories involving warring nations.

To make a long, disturbing, and shitty story short, the British government chemically castrated Turing for being gay and (maybe, probably) poisoned him for fear of having their secrets revealed. What a great *thank you for your service*. Fucking hell.

While Turing had to endure all that shit, he quietly contributed some of the most important advances in human knowledge. In standard patriarchal terms, Turing is said to be the "father" of computer science *and* artificial intelligence. That's a lot of fucking descendants. If there is anything to the idea of posthumous justice in the world, Turing has it. The most important international award in computer science is named after him, for example. And thousands of researchers followed his footsteps in asking deep questions about the limits of computation.

It took a long time, but finally in the late 1970s, people started to think about how the laws of physics impact the ability to compute. Real computers use energy, and energy is the subject of physics after all. Naturally, this led to the road of quantum physics. Then, in a truly miraculous turn of events, David Deutsch thought about how computation can be done in the parallel worlds of the multiverse, and quantum computation was born. Of course, it's more subtle than that, but this is not a fucking history lesson. Google it, for fuck's sake. In any case, no one ever said crazy ideas can't lead to useful outcomes, but that is not a recommendation. So let's ignore this historical quirk.

Remember that quantum things like atoms have discrete energy levels. The smallest number of levels you could imagine is two. If you label those levels as "zero" and "one," you have something quite similar to the smallest unit of information in Turing's computers, which is called the *bit*, by the way. (Other historical quirks have us favor the *byte*, which is eight bits.) These two-level quantum bits are now called *qubits*, and a hypothetical machine built of qubits is called a quantum computer.

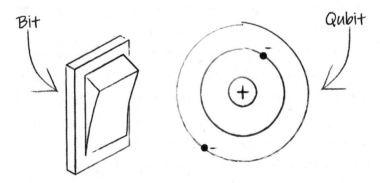

Just before the turn of the twenty-first century, scientists discovered that programs could be created for these quantum computers that solved problems faster and with fewer qubits than bits in a digital computer. This promised a fundamentally new way to solve *some* computational problems. Since then, we've been trying to build these machines. Progress may seem slow for the kind of people that are all like, "you promised me flying cars fifty years ago too!" But there is nothing fundamentally stopping quantum computers from being built at scale. We're just missing the usual ingredients: money and people. I'd prefer the former if you have it.

But Wait! There's Even More

The qubit not only defined a new type of computer but also gave us a new kind of *information*. A qubit, the smallest amount of information that can represent the state of quantum things, can be a superposition state, and many of them can represent entangled states. These are features in quantum mathematics that regular old ones and zeros cannot possess. Even the seeming obstacle of quantum uncertainty has novel applications. Recall from chapter 3 we discussed *quantum cryptography*. Using quantum information, we can have perfectly secure communication. There are even commercial quantum cryptography systems available today. Go get your quantum on!

The very fragility of superposition and entanglement have applications as well. It's kind of obvious, really. If you want to detect a tiny effect, you need something really sensitive. So-called *quantum sensors* are being developed for this purpose to detect tiny changes in, especially, magnetic fields. The principle is not unlike MRI, just on a smaller, nanoscopic scale.

These twenty-first-century quantum technologies—quantum tech 2.0, you might call them—of course use physical quantum systems. But so do digital computers and other twentieth-century technologies. The difference is in what *information* they send or process. Digital technology breaks down information into bits—ones and zeros—which can only be flipped back and forth. Quantum technology uses native quantum states as information, qubits.

> For the last hundred years, technological *hardware* has been based on our understanding of quantum physics. For the next hundred years, technological *software* will be based on our understanding of quantum mathematics.

Laser Eyes

Have you ever seen a diagram of how sight is supposed to work? They usually have an arrow pointing from an eye to an object. Of course, nothing is coming out of your eyes—unless you are watching the news, in which case tears are coming out of your eyes, or you are fucking Superman. In case you've been living under a rock for the past several decades, there is a lot of science fiction and fantasy revolving around the idea of lasers as weapons. Sometimes the lasers come out of an alien's eyes and other times out of big fucking guns. But two things are always true: you can see them as glowing streaks, and they sound like this:

pew, pew, pew. Okay, that doesn't really work in this medium, but go ahead and google "laser sounds," and you'll get the gist. Lasers don't actually make sounds, but no one is going to watch a silent action movie. More importantly, you can't actually see lasers.

That's right, you can't see a laser beam. Well, not unless you are looking directly at it, which I don't recommend. To *see* something means to have light bounce off it directly into your eye. Light doesn't bounce off light. Grab your laser pointer, and point it anywhere. Can you see it? Not unless it hits something. Whether you are entertaining a cat or the audience of a presentation (they're not much different), the subject *sees* the laser as a spot on the wall or screen. They see the red dot because the laser light hit the wall and was scattered in all directions. No one sees the beam of the laser; it's just not possible. The laser light is traveling along a straight line in a direction that shouldn't be level with your eyes. One of you is going to shine a fucking laser in your eyes, I just know it. Please don't contact me if it's you.

While it is true that laser light can scatter off air molecules, those are few and far between as far as a photon is concerned. If you point a laser through a cloud of smoke, however, you can see light being scattered along the straight-line path of the laser beam. In a sense, this shows you the beam. This effect is put to good use in rock concerts and when I enter my classroom for lecture—yeah, I'm that badass. (No, I'm not. I deliver my lectures online while not wearing pants. Sad emoji.) But there is no smoke in outer space, so the idea that you can see and hear space laser weapons just doesn't jive with quantum physics. Speaking of which, the whole idea of laser weapons is pretty stupid given the existence of, you know, mirrors. You'd think that technology over one hundred years old would be immune to such misconceptions. On the other hand, quantum physics totally ruined *Star Wars* for me.

Cosmic Rays and Spider Bites

The origin stories of many superheroes (and Disney princesses) involve losing their parents. And if that wasn't bad enough, they then have their DNA changed by some further catastrophic event. (The superheroes, not the princesses.) Sometimes, it's gamma rays from a nuclear meltdown. Other times, it's cosmic rays from some cataclysmic space explosion. There have even been superheroes created with radioactive spider bites. Those are actually pretty common in Australia though.

While radiation can cause small, localized mutations of your DNA, it's more likely to lead to cancer than it is to allow you to break the laws of physics. But that is no reason to fear the word *nuclear*. We already saw way back in chapter 1 how quantum energy works. I told you then about how your own body is radioactive. What sort of nuclear radiation should you be worried about then? The answer is the result of tedious calculations of probabilities and risk. Scientists and engineers have done the math. I assure you, your busybody activist friend who gets their information from internet memes has not done the math. Go ahead and get that 5G phone.

Quantum Qompanies

This is my favorite thing about future technology—no one can protest its use after falling for some conspiracy theory about it.

No one is out there with signs that say, "They're putting quantum computers in our schools! Those things use mind-controlling waves to poison children with leftist ideology, and they may cause cancer, but more importantly, the hell I'll let my kid turn into a dirty lib!" Not yet anyway.

Don't get me wrong. There is bullshit about quantum technology, but this time, it is from the entrepreneurial posers who latch onto anything that looks new and shiny with the word *tech* in it. Hop on LinkedIn or some other "networking" site if you don't know what I'm talking about. The embarrassing call to action goes from "buy my quantum product" to "invest in my quantum company." It's much worse in a way. At least with woo peddlers, you have some cheap pieces of plastic you can enjoy. With quantum entrepreneurs, you don't get anything but empty promises. It's kind of like voting for political parties.

To be fair though, even here in 2022, there are plenty of legitimate quantum technology companies. Don't know of one? Look for the Q where it shouldn't be—every company dealing in quantum technology needs a Q in its name or product. Of course, not being a quantum physicist, you can't tell which ones are honest. Don't worry. I have a quantum start-up company that can help you with that—Quantum Qonsulting. I just need you to invest $10,000 in Best Buy gift cards.

I hope by now you'll be able to spot the red flags. If your potential investment opportunity is built on quantum fluff, you are in trouble. Quantum energy, quantum superposition, quantum entanglement, and other quantum buzzwords don't make for a profitable business plan. Your grandmother could have told you this, but if it sounds too good to be true, it fucking is!

You know it, Granny. Save your money or spend it on books—I know an author who writes way too many of them.

Analogies, Metaphors, and Similes

A simile is like a metaphor not used as an analogy, or something like that—I had to google the difference again. But I think I got it. "Life is *like* a box of chocolates." That's a simile. "Life *is* a box of chocolates." That's a metaphor. "Life is like a box of chocolates. You open it up only to find a disappointingly melted goo of separated canola oil and cocoa-flavored syrup." That's an analogy.

Time for your quiz.

Here's something you'll see a lot in tech and business magazines—a bit can be zero or one, but a qubit can be zero *and* one at the same time. What is this?

(a) simile
(b) metaphor
(c) analogy

You got it! The answer was indeed (d) bullshit. It's not a simile. (It doesn't use the words *like* or *as*.) It's more like a metaphor struggling to be an analogy. It's a bad analogy. It's bullshit. Making a bad analogy at such a low level of description is a recipe for disaster. Every explanation it attempts to provide compounds the nonsense. For example, people take the idea of a qubit being zero and one at the same time to mean that a quantum computer can check every solution to a problem at once. That's not technology—that's fucking magic.

A qubit is not zero and one at the same time. It's *kind of like* a bit that can be zero and one at the same time, but it's not literally that. When in doubt, hedge with a simile and end it there. But why does the metaphor fail? It's really quite simple. The zero and one of the bit are labels for any two mutually exclusive things. Mutually exclusive means one thing or the other but not both. True or false. Those are mutually exclusive labels. Something can't be both true *and* false. So the metaphor is really saying a qubit is equivalent to logically contradictory bullshit. This is probably not the best analogy to go with to make the case for breakthrough technology.

A qubit, as alluded to, is the state of the smallest type of quantum system, like *spin* for example. It can be written numerically but requires more than a zero or one. But that doesn't sound sexy enough for venture capitalists. I guess that's why no one is investing millions in my quantum start-up. I really should have gone with the quantum parallel universe Ponzi scheme.

Teleport Me Out of Here

Teleportation was made famous by *Star Trek*'s "Beam me up, Scotty!" The person would disappear from some distant location and reappear onboard in the transporter room. The idea of magically changing locations definitely goes back before any scientific mechanism could be claimed to be responsible for it. Now, it seems, people can legitimately claim quantum physics is the science of teleportation because someone went and called an actually important quantum communication protocol *quantum teleportation*. So unfortunately, we must say that *quantum* teleportation is real. And every fucking time we do, we have to also say that it is not like teleportation in *Star Trek*.

Suppose I have a qubit of information I want to send to you. I'm a nice guy like that. But it's going to be difficult, especially in this day and age, since modern communication technology was built to send *bits* around, not qubits. There is a way I can do it, though, provided we already shared a pair of entangled qubits. (We almost certainly did not, but we can pretend.) The procedure requires me to send you two bits of data and you to perform some calculation with your half of the entangled pair. After that, you possess the qubit of information. No quantum information traveled from me to you, yet the information appeared at your location. That sounded to the group of researchers that discovered the procedure in 1993 like the science fiction version. So probably tongue-in-cheek, they called it quantum teleportation. Lesson learned—don't name your science ideas with clever puns if they are going to end up being super important. Actually, I probably don't have to worry about that.

Ultimately, none of this matters if you just go about your life with a strong spam filter, because product engineers don't want to confuse you. They want the opposite. They want "plug and play." Every second a potential user has to spend thinking is a second wasted. The next revolutionary quantum technology, when it arrives, will be invisible to you.

Quantum Inside

Technology development is a complex process involving even psychologists who aim to understand how it will be used. Luckily, there's no such thing as "quantum psychology." (Please, let there be no such thing as quantum psychology.) So there are always parts of technology that incorporate our earliest inventions, such as language and writing. Even if Grandpa doesn't want an iPhone—I don't blame him—he could still *use* it. You could raise Henry Ford from the grave and sit him in a fucking Tesla, and he'd happily drive away.

Of course, behind the interface of the car is technology beyond his wildest imagination. Yet there will always be a way to explain it because the ultimate purpose it serves is for humans. Humans happen to also be the people who buy books that explain stuff. I've tried to explain quantum physics and technology from your perspective, fellow human. If some future technology is built solely for robots to use, it won't need to be explained. I won't write books for robots, and robots won't either. Unless they are willing to pay—I'm just a scientific shill after all.

8

Where the hell do
I go from here?

What the fuck just happened? Well, I kind of tricked you. I told you that I wasn't going to tell you what quantum physics *is* but what it *isn't* by dispelling all sorts of bullshit. We did that together—give yourself a pat on the back. As you pat yourself on the back—you know, feeling good about yourself—you may get a sense that you actually do know some things about what quantum physics *is*. Congrats. But don't get too cocky, or you'll feature in my next book. Also, check for moles back there—quantum physics can be a real bitch.

Defining things is hard. Sometimes knowing what something *is* is just knowing all the things it isn't. So you have indeed learned quite a bit in these pages about quantum physics—enough, anyway, to save yourself a few dollars and a lot of embarrassment. I'm not sure if you could tell, but this book was an attempt at humor. But it wasn't the kind of pointless humor you find in Adam Sandler movies or intentionally offensive humor you find as memes bleeding out of every internet subculture forum. The shit I said about quantum physics was factual.

I really couldn't help myself. Some people pay me to teach this stuff after all.

You know what, I'm just going to confess. The point was to teach you about quantum physics all along. I told you about the most important concepts in quantum physics and roughly in chronological order too. What a twist! It was really a five-star performance. This book probably deserves one of those shiny silver emblems on the cover with the imprint of some dead person's face that no one recognizes.

Quantum Shit You Now Know

If you aren't feeling as enlightened as I'm assuming you should be, here's a quick recap.

Quantum energy comes in chunks—called quanta—which set the structure of matter, and it fingerprints the light it emits. This is actually a restriction on the classical idea of energy. You wouldn't want quantum energy if you had the option. But you don't have the option. All is quantum, including ionizing radiation, which you don't want in large doses. Luckily, scientists understand this well, and standards and regulations protect anyone who might be exposed.

Both matter and energy display wave-like or particle-like behavior depending on the context. This is called wave-particle duality. The larger something is, the more subtle the wave effects will be. Even for something the size of a bacterium, the wave effects would be nearly impossible to capture. So the idea that *you* have a quantum wave nature is correct but irrelevant.

Quantum uncertainty means that not everything that *could be* known about an object *can be* known about it. The prototypical example is position and speed—you can't measure both the position and speed of an object to perfect precision. In fact, it's a bit stronger than that. Position and speed just aren't perfectly defined properties any object *has*. Again, for large objects, "perfect" is not really relevant if it means better than one part in a trillion-trillion-trillionth of the size of an atom, which is definitely good enough for any purpose.

Quantum superposition is the idea that states—mathematical descriptions—of objects can be added together to obtain other valid states. This is obvious from the wave-like nature of quantum physics. Two waves can be added together to get a new wave. What happens when a wave hits the shore? It breaks up into a chaotic mess. It takes a lot of work—an impossible amount probably—to maintain the purity of quantum waves for even well-isolated microscopic things. If someone claims superposition is responsible for something large enough to be visible, they are lying.

And there was quantum entanglement, parallel universes, and quantum fucking computers. Those were only a few pages back. I'm not going to do everything for you.

All the Shit You Should Avoid

This book was about quantum physics but also bullshit. There is nothing particularly special about *quantum* bullshit—I'm not going to go inventing the word *quantumshit* or anything like that. Most bullshitters probably can't even distinguish that they are using terminology from quantum physics instead of just technical-sounding words. It's bullshit that is the real problem—quantum physics just gives the bullshitters more fuel.

So what is bullshit anyway? Hmm...perhaps this should have gone in the introduction. Fuck it—better late than never. But sometimes, defining things really drains the excitement out of it. This is especially true for value judgments, which always have a subjective element to them. One is reminded of Supreme Court Justice Potter Stewart's description of *obscenity*, which can be paraphrased as, "I'm not going to define 'hardcore pornography,' but I know it when I see it." Surely the judge would just rather see it than sit around not seeing it during some boring academic discussion of it. Likewise, we may not have an airtight definition of bullshit, but we know bullshit when we see it. My task was only to upgrade your existing bullshit detector for the quantum age.

The problem in defining bullshit is that what counts as bullshit partly relies on the *intentions* of the bullshitter. But relying on this fact gives us a succinct working definition—bullshit is deceptive nontruth. It's not necessarily a lie, because a lie implies the liar knows the truth. A bullshitter just doesn't care. A quantum bullshitter almost certainly doesn't know what is *true* about quantum physics, so they can hardly lie about it. They just use quantum jargon because they think they sound more convincing that way.

It's not all about deception either, because otherwise, we'd have to call cats bullshitters too. You never wanted my affection. You just wanted my food—damn you, Mittens! Bullshit is uniquely human. Sure, animals deceive predators and prey, but this is instinct. Humans thoughtfully lie, cheat, and steal because—unlike cats—they all have something to sell you. We're not buying it. What? Those Shake Weights? Those aren't mine. The previous owner left them here, I swear!

All the Other Shit

You may also wonder, *why bullshit? Why not horseshit?* As it turns out, I probably could have replaced some bulls with horses, but it does have a slightly different meaning. It's easiest to think about what these terms *imply* as accusations. Bullshit implies deception. Horseshit is also nonsense but stems from ignorance instead of deception. Some random well-meaning person might be keen to "explain" that they just learned cats can be dead and alive at the same time. That's horseshit, not bullshit—unless you are asked to click something.

You may be familiar with *Occam's razor*, which is usually summarized as the adage *the simplest explanation is the best one.* It's a "razor" in the prescriptive sense that you should "cut away" as much of your explanation as possible so long as it still works. Occam's razor is often applied to shoot down conspiracy theories. The Earth looks and feels flat to me. Is that because it really is flat and there is a massive international conspiracy including twenty-five million actors claiming to live on an imaginary continent called "Australia," or is it because the Earth is a really fucking big ball and I am really small? Occam gives you the answer.

There's another, more relevant "razor" called *Hanlon's razor*, summarized as *never attribute to malice that which is adequately explained by stupidity.* In other words, don't mistake horseshit for bullshit. Bullshit is a serious accusation. Before you *call* bullshit, you better be certain you have a good case for your claim. Lucky for me, I have a PhD in quantum physics and can just wield my authority around like I'm Tom Cruise in an amateur acting class—or Scientology induction. Fair? Probably not. But I didn't waste twelve extra years in school for nothin'. When you call bullshit, you are putting *yourself* in a vulnerable situation. I don't call bullshit all willy-nilly on everything, just the quantum crap. If I call bullshit on anything else, it's because I've fact-checked it. You can do the same by bookmarking many nonpartisan fact-checking websites. But when in doubt, you are better off yelling horseshit instead of bullshit. Actually, maybe just don't yell shit in public? Nah. Don't be a chickenshit—let's go apeshit!

If You Fight Bullshit...

You probably know the punchline already. If you fight shit, you are going to end up covered in shit. It's really as pointless as it sounds. In fact, before you get all self-righteous about it, better advice than to be carefully calling bullshit is to just fucking ignore it. Since we are all about razors now, this one is called *Hitchens's razor,* and it goes like this—*what can be asserted without evidence can also be dismissed without evidence.* Christopher Hitchens used this in the context of debates on religion, but it equally applies to bullshit.

If someone comes at you with dubious, sensational, or otherwise questionable claims and insufficient evidence, it's on *them* to go and get more evidence, not on *you* to find arguments against their claims. You don't need to prove that salt crystals aren't powered by quantum energy; *they* need to show *you* the evidence from well-regulated clinical trials. In fact, you don't even need to call bullshit at all. What you should do is kick them in the shins and ru... Kidding! But you could tell them politely to fuck off.

Is this too pessimistic? No. Should we really be just ignoring anything we don't want to listen to, sloughing it off as bullshit? Yes. You have a finite amount of time on this planet. You could spend the entirety of it wading through bullshit, but that's no way to live. The reality is that dispelling bullshit is an impossible uphill battle. The amount of time needed to refute bullshit is ten times longer than it took to make it. It's the shittiest game of whack-a-mole ever.

Take this very book as a cautionary tale. I've spent fifty thousand words on quantum bullshit and barely scratched the surface. Hopefully, at least one reader has been spared some embarrassment. But more likely, all I've achieved is enticing one of you to shine a fucking laser in your eyes.

You Can Choose Your Friends but Not Your Bullshitters

Oh, you are going to ignore my advice and engage anyway? Fine. I guess, in some cases, it really can't be avoided. This might be especially true if the bullshitter happens to be a family member who just discovered Facebook and thinks only facts are allowed to be posted on the internet. There is a crucial distinction to be made between bullshitters and people who *fall for* bullshit and end up repeating it. If Grandpa says that he read on the inter-tubes that vaccines have radiation in them, that's horseshit, not bullshit—unless of course Poppy has a side hustle selling alternative medicine out of the back of the old-age home.

Do people deserve ridicule? No. And yes. The two most common themes in insults are stupidity and depravity. In the former case, we have words like *moron, dumb, brainless, idiot,* and so on, while, in the latter case, we have words like *deplorable, disgraceful,* and *shameful.* I'm sure it's not a coincidence that the latter case always requires more syllables.

It's important not to ridicule people who *believe in* bullshit. Regardless of what you think about insults, and far be it for some asshole quantum physicist to lecture you about morality, the fact is, you can rarely change behavior by insulting someone's intelligence. There is ignorance, and there is *willful* ignorance. Naive people who fall for bullshit deserve your sympathy, not your scorn. Go on, insult those amoral bullshitters all you want—you only want them to go away after all—but don't insult Papa.

Live Your Best Quantum Life

It's kind of sad, really. Quantum bullshitters—in fact, all bullshitters—are preying on the vulnerable. The ones most likely to fall for it are the most desperate. I'm afraid quantum physics really does have nothing to offer those people. Sorry. But what about the other people? The lost, the fallen, the sad, the lonely... you, perhaps? Can quantum physics help you, if only in some small way? Maybe. Maybe, by taking some lessons from quantum physics, we can avoid a midlife crisis impulse or something. It couldn't hurt in any case. I'm not aware of quantum physics ruining any marriages...yet.

Quantum uncertainty reveals that the world has a fundamental level of unpredictability to it. Sure, you're not going to go to the laboratory and prove this yourself, but perhaps there is a hint of quantum there worth sprinkling into your worldview. The world is complicated. Relationships and financial matters are complex. Even individual people are unpredictable. Does it have something to do with quantum theory? Probably not, but the moral of the quantum story holds—you are never going to get rid of it. Like those persistent and determined quantum physicists of yore, you just have to fucking deal with it.

Quantum theory demonstrates that a rigid, objective world does not exist independently of the way we choose to interact with it. What a quantum physicist chooses to measure in the lab matters. What you choose matters. There is no path down which everything is easy, and all choices are equal. While that fantasy might absolve you of potential guilt, it would remove the one thing that makes you human, that quantum physics grants you—agency. Go on, make your world. Just watch out for the bullshitters making their own.

Endnotes

Chapter 2: Fucking matter waves

i   Ben Zimmer, "'Resonate'." *New York Times*, November 18, 2010. https://www.nytimes.com/2010/11/21/magazine/21FOB-onlanguage-t.html.

ii   "The Crystal Bible, Volume 3," December 8, 2021. https://www.octopusbooks.co.uk/titles/judy-hall/the-crystal-bible-volume-3/9781841814605/.

iii   Laura Berman, *Quantum Love: Use Your Body's Atomic Energy to Create the Relationship You Desire* (Carlsbad, CA: Hay House, 2017).

iv   Ledyard King, "Do Wind Farms Cause Cancer? Some Claims Trump Made about the Industry Are Just Hot Air," *USA Today*, April 3, 2019, https://www.usatoday.com/story/news/politics/2019/04/03/cancer-causing-wind-turbines-president-donald-trump-claim-blown-away/3352175002/.

v   J. M. Randall, R. T. Matthews, and M. A. Stiles, "Resonant Frequencies of Standing Humans," *Ergonomics* 40, no. 9 (1997): 879–86, https://www.tandfonline.com/doi/abs/10.1080/001401397187711.

Chapter 3: We have no fucking clue what is going on

i   Werner Heisenberg, "Über den anschaulichen Inhalt der quantentheoretischen Kinematik und Mechanik," *Z. Physik* 43 (1927): 172–98, https://doi.org/10.1007/BF01397280.

ii   Robert P. Crease, "Quantum of Culture," *Physics World*, September 1, 2008, https://physicsworld.com/a/quantum-of-culture/.

iii   Robert P. Crease and Alfred Scharff Goldhaber, *The Quantum Moment:*

How Planck, Bohr, Einstein, and Heisenberg Taught Us to Love Uncertainty (New York: W. W. Norton, 2014).

iv   *Numb3rs*, episode 2, "Uncertainty Principle," directed by Davis Guggenheim, written by Nicolas Falacci and Cheryl Heuton, aired January 28, 2005, on CBS, https://www.imdb.com/title/tt0663233/.

Chapter 4: That fucking zombie cat

i   "Volume 15: The Berlin Years: Writings & Correspondence, June 1925–May 1927 Page 566." The Trustees of Princeton University. Accessed April 1, 2022. https://einsteinpapers.press.princeton.edu/vol15-doc/678.

ii   Paul A Dirac, *The Principles of Quantum Mechanics* (Oxford: Clarendon Press, 1930).

iii   *Merriam-Webster.com Dictionary*, s.v. "superposition," accessed February 8, 2022, https://www.merriam-webster.com/dictionary/superposition.

iv   *Numb3rs*, episode 8, "Identity Crisis," directed by Martha Mitchell, written by Wendy West, aired April 1, 2005, on CBS, https://www.imdb.com/title/tt0663217/.

v   Wikipedia, s.v. "Goop," accessed February 9, 2022, https://en.wikipedia.org/wiki/Goop_(company).

vi   "The Energy Experience," *The Goop Lab*, hosted by Gwyneth Paltrow and Elise Loehnen. Aired January 24, 2020, on Netflix.

Chapter 5: Faster than fucking light

i   Aloys Leo Prinz, "Chocolate Consumption and Noble Laureates," *Social Sciences & Humanities Open* 2, no. 1 (2020): 10082.

ii   Albert Einstein, et al., *The Born-Einstein Letters* (London: Macmillan, 1971), 158.

iii   E. Schrödinger, "Discussion of Probability Relations between Separated Systems," *Mathematical Proceedings of the Cambridge Philosophical Society* 31, no. 4 (October 1935): 555–63.

iv   Jeffrey Kluger, "What Einstein Got Wrong about the Speed of Light," *Time*, October 22, 2015, https://time.com/4083823/einstein-entanglement-quantum/.

v   Gabriel Popkin, "Einstein's 'Spooky Action at a Distance' Spotted in Objects Almost Big Enough to See," *Science*, April 25, 2018, https://www.sciencemag.org/news/2018/04/einstein-s-spooky-action-distance-spotted-objects-almost-big-enough-see.

vi   ABC Science, "Einstein Was Wrong: First Conclusive Proof 'Spooky' Quantum Entanglement Is Real," ABC News, October 22, 2015, https://

www.abc.net.au/news/science/2015-10-22/einstein-was-wrong3a-spooky-entanglement-is-real/6876262.

vii   John Markoff, "Sorry, Einstein. Quantum Study Suggests 'Spooky Action' Is Real," *New York Times*, October 21, 2015, https://www.nytimes.com/2015/10/22/science/quantum-theory-experiment-said-to-prove-spooky-interactions.html.

viii   Clara Moskowitz, "Quantum Entanglement Creates New State of Matter," *Scientific American*, September 22, 2014, https://www.scientificamerican.com/article/quantum-entanglement-creates-new-state-of-matter1/.

ix   Eric Szeto, Asha Tomlinson, and Virginia Smart, "'This Is Snake Oil': Scientists Don't Buy Balance-Boosting Clips Featured on Dragons' Den," CBC News, February 2, 2018, https://www.cbc.ca/news/business/quantum-wellness-clips-marketplace-1.4513382.

Chapter 6: Infinitely many goddamn worlds

i   Stefano Osnaghi, Fabio Freitas, and Oliver Freire Jr., "The Origin of the Everettian Heresy," *Studies in History and Philosophy of Modern Physics* 40, no 2: 97–123.

Check out a sneak peek of
Chris Ferrie's upcoming book,

42 Reasons
to
Hate
the
Universe,

publishing in February 2024.

Why hate the Universe?

Twinkle twinkle little star,
How I wonder what you are...

We were all sung this as children as we gazed up to the heavens at the tiny points of light in the night sky in wonder and awe at the beauty and mystery of the Universe. Of course, today we wonder less about the stars. Science has given us a new lens on the Universe, allowing us to see beyond what only our eyes can tell us. We have so much more to learn about the Universe, which seems to be a place of infinite possibilities. It is a place of wonder. It is a place of mystery. It is a place of beauty, love, and hope. And, although science has taken us on a journey beyond the stars, these twinkling specs of light are where it all began.

Your parents probably told you that the stars were big balls of burning gas—which is wrong—and that there are millions of other stars in the Universe, but *our* star, the Sun, is special—which it isn't—and that special star shines just for you because you too are special—which you aren't.

There *is* beauty in the world, though, or so it would seem. Our brain surges with endorphins—the "feel-good" chemicals— when we see a glowing sunset, catch a glimpse of a shooting star, or receive our first kiss. Optimistically, we want to believe the Universe made these moments especially for us. Alas, it is an illusion. The Universe cares about you but only inasmuch as it wants you dead. Once you take the rose-colored glasses off, you soon realize that the Universe really wants nothing to do with you; often it is trying to kill you. To see the world for what it really is—the naked truth—is granted by the power of science. But what does science say exactly about our existence or, more relevantly for this book, our journey toward death?

Consider that you are just a group of atoms structured in a specific way for just long enough that you can try to understand this thing we call existence. These atoms were formed in the hearts of massive, long-gone stars and thrown out into the cosmos when the star died in an immense explosion. These atoms likely drifted between countless forms—making up parts of nebulae, asteroids, or perhaps even another planet—before finding their way to our solar system, and then the Earth, before becoming a part of you.

But these atoms are not special and could have just as easily been used to make the dog shit you are cleaning out of your shoe treads. The idyllic and poetic vision of us as divine machinations of the Universe is a lie. The Universe wrote the rules to the game of survival, and the second rule says nobody wins (we don't talk about the other rules). This rule is called decay, and it says that you, we, *everything* will rot away into dust and spread thinly throughout a cold, dead universe. But don't worry, there are plenty of ways for the Universe to get you before that happens.

Take water, for example—beautiful, picturesque, music to the ears, and of course, essential to life. Surely water can only be considered a precious gift from the Universe. Ripples on a pond, waves crashing on white sandy beaches, a cool mist on a hot day, or a hot shower on a cold day. Turns out most of it will kill you, or at least make you fall violently ill. Then what about the source of all our energy, the Sun? It has been worshipped for millennia and is the basis for all of our food here on Earth. It can only be good for us, right? Well, no. While it is responsible for producing the delicious burrito you bought to eat while reading this book on your lunch break, it is also responsible for most skin cancers. So it seems even the most indispensable gifts from the Universe come with serious caveats.

This all might be more palatable if we had some other intelligent beings to share our plight with. We've called out into the cosmos, searching for signs of life. Sure, our message included some tasteful nudes that could discourage a reply, but maybe there is just no one out there. Maybe the Universe is just that cruel to drop us in a typical galaxy, next to this unremarkable star, on a lonely planet forever on the precipice of annihilation.

So what has the Universe placed out in the cosmos other than intelligent life? Not much, actually. We're essentially alone and trapped in the tiniest corner of a void we wouldn't survive in even if we could escape our pale blue dot of a prison cell. Either we'd end up too far from any star and freeze to death, or if we get too close, be irradiated with electromagnetic fire. What we can't see—beyond the stars and galaxies that have earned our worship—is terrifying beyond imagination. Amongst the beauty of the constellations and cosmic dust are black holes that can rip us apart with their immense gravity or dark matter particles that would burn a basketball-sized hole through our body.

Of course, it would be inhumanly cynical to say that the Universe is not amazing, awe-inspiring, and beautifully complex—but that's not why you bought this book, is it? You purchased this book because you wanted to know what an asshole the Universe is, and believe us, you have no idea what else it has in store for you. The Universe is dark, cold, and cruel. So strap in and get ready for the most soberingly uncomfortable take on the Universe since *A Brief History of Time*.

If you are looking for inspirational quotes to tell your friends at a party, post on Instagram, or make you sound like an intellectual, you are better off purchasing a thesaurus. If you are looking for cynical rants and justification for how unfair your life is, you won't find that here either—go listen to a podcast. This book's goal is to tell the somber tale of what science has uncovered—a darker and more disquieting side of the Universe.

Oh, and if you *are* the Universe, the jig is up—we are calling you out. From extinctions to murderous artificial intelligence, all the way to the collapsing of space and time itself—this is bullshit! Sure, many of the shitty things you have in store for us are preventable, and there is a perverted sort of beauty in staring into the darkness of your infinite depths. You may have fooled people with your pretty space dust and golden ratios, but these things are just a front. We have science now, and we are watching you.

Reason 1

No one else has made it this far

Once you look past the drugs, self-righteousness, vanity, and physical and psychological abuse, the Oscars could somewhat be considered a celebration of talent. When winners aren't climbing the stage to assault the host, they are accepting awards for their ability in pretending they are something that they are not. They stand there in all their flawlessness, with clothing worth thousands, only to tell the rest of us to *follow our dreams* so that we too can one day attend an event filled with the pinnacle of what humanity has to offer.

What they fail to mention is that regardless of whether we *follow our dreams* or not, the absolute vast majority of us will never be as successful as they are. And this isn't just for actors—it's true also for sports stars, writers, artists, or pretty much any profession that enables you to have sex with people who otherwise wouldn't usually want to have sex with you. Alas, it appears as if there is some sort of filter to ensure that people like us will most certainly never reach that level. We may not look right, talk a certain way, or perhaps lack the ability to be a pretentious, self-righteous prick. Now let's take this analogy about celebrities and apply it to something less meaningful, like the long-term survival of our civilization.

Developed by economist Robin Hanson, the *Great Filter* is an idea that attempts to explain why we do not see civilizations—or life for that matter—anywhere else in the cosmos (it's bad no matter which way you swing it—see Reason 29). The Great Filter is a hypothetical barrier that prevents the development of technologically advanced species. It's necessary since otherwise, by now, we would have expected to be overwhelmed by intergalactic junk mail. What we don't know is if this barrier was in the past and we overcame it, or if it is a barrier our civilization is yet to face. The Great Filter is essentially a resolution to the famous *Fermi Paradox*, which points out that we have still not discovered life in spite of the fact that there are billions of other solar systems similar to ours out there.

To fully understand the rarity of life in the Universe, we must first understand how hard it is for life to form. There are major challenges that lay between having a planet suitable for hosting life to the evolution of a life form capable of creating technosignatures (a fancy word you can read more about in Reason 29... if you make it that far).

So for humanity to have gotten this far *may* indicate that we have overcome the challenges that would otherwise limit life from forming elsewhere in the Universe. We have a suitable star and planet system that enabled life to originate in the first place and also enabled the transitioning from simple single-celled organisms to more complex organisms with inner working parts. Our pocket of the Universe has also allowed for life to develop sexual reproduction—albeit many still struggle with this—and as such house multicellular organisms with some degree of intelligence that are capable of using tools and mobile apps.

According to Hanson, humanity currently sits at the second-highest level in the nine-step process of an evolutionary path. This means that our civilization is advancing toward the highest level—a *colonization explosion*. A colonization explosion is where we spread ourselves through the solar system and eventually the galaxy. We already possess the technologies to send robots to inhabit other planets, so it's amazing we can't be bothered sending the equivalent weight of eight blue whales to get a human there (that'll make sense when you get to Reason 26). Just like the brave voyagers before us, future generations of billionaires are frothing at the chance to take selfies in all areas of our galaxy, especially if it means the possibility of increasing their Twitter followers.

Let me paint a picture for you. Imagine that we did find life out there in the cosmos. The detection of extraterrestrial life would hail as the greatest discovery in the history of humanity. It would provide an answer to one of the most profound questions ever asked in a British accent (posh, not cockney) over drone footage of waves crashing on a rocky shore as the camera pans past a sunset to a shot full of darkness instilling an existential thrill in our hearts, and it resonates... *Are we alone?* And here you thought the most profound question was how old those left-over chicken wings in your fridge were. Anyway, when we look out into the cosmos for intelligent life we find, well, nothing. So apparently, we are alone. Also, the chicken wings have probably turned. Things are just going from bad to worse.

Believe it or not, the absence of sentient beings out there could very well be a good thing. Finding extinct alien civilizations that were more advanced than our own could indicate that the Great Filter is ahead of us, and any number of very bad things lay waiting for humanity. To be fair, this could also be regarded as good news for people on Twitter who only post GIFs of dumpster fires.

Let's consider a number of events that are possible, even likely, to occur in our lifetime and what they could mean. Although it might be hard for many of us non-Oscar-winning simpletons, we are going to ask you to focus on something other than yourself for a moment. Imagine that humanity has discovered what many believe to be the most likely form of alien life—a basic single-celled organism that lives on Jupiter's moon Europa. Where would that put humanity in the Universe's grand plan? Counterintuitively, this plausible—albeit fictitious—discovery of a rare yet very simple life form on Europa would suggest that it is incredibly *difficult* for life to transition past the simple cellular phase.

Consider the *endosymbiotic hypothesis*. It suggests that about 3 billion years ago, one single-celled organism "swallowed" another particular single-celled organism, but instead of becoming lunch, the two cells figured out how to work together. Scientists believe this is the original love story that eventually led to the mitochondria and other less-internet-famous organelles. And those scientists wear lab coats, so you know you can trust them.

Both the endosymbiotic hypothesis and discovery of simple life on Europa, or elsewhere in our Solar System for that matter, would insinuate that the filter is behind us. This would be great news, as we could very well be masters of our own destiny. Some go further and believe the very best outcome in our search for E.T. is to find nothing at all. Albeit rather lackluster, this could indicate—that by sheer luck—life on our planet has managed to break through the Great Filter. This event in our evolutionary past could have been so improbable and rare that it allowed us to overcome a challenge that many other potential life-forms in our Universe have not.

Or perhaps the Great Filter is still ahead of us. The Universe is one gigantic beast, and it would be ignorant to think we have searched thoroughly enough to conclusively determine that alien life is indeed as rare as we currently claim. In any case, if there once were intelligent beings in the cosmos, they appear to be extinct—that, or they've moved far away from this shithole of a Universe. Hopefully they have internet memes there. But who really knows—they could've been swallowed up by a black hole, eviscerated by a supernova, overwhelmed by disease, eradicated by their own technology, or maybe just threw their tentacles up in the air before sending their spirits out into the cosmos to embody other beings. Although, if that last part were true, it would mean a few of those celebrities with faces prettier than their brains might be onto something.

Be that as it may, the very fact that we have gone through an evolutionary path from simple single-celled organisms to complex primates—with an unhealthy obsession for other primates who are good at pretending to be other primates—is an amazing achievement. That, or it was some fluke wherein a brainless organism accidentally ate something that made it better at reproducing. Probably don't mention that part in your acceptance speech though.

Reason 2

We are programmed to be selfish pricks

In this chapter, we discuss some of the most treacherous holes in the Universe. No, not black holes (you can see Reason 40 for those), nor are we talking about the dark spot on the surface of Uranus, although it's not far off. We are talking about assholes, and by assholes, we mean you. We know, we know, you're thinking *Pfft...you don't know me*. I [insert justification as to why you are not an asshole]! But hold up. We're not saying being an asshole is a bad thing. In fact, we would argue it could even be a good thing—if you enjoy stuff like being successful, building relationships, finding a partner, and, well, not dying.

So what does being an asshole have to do with the Universe? Well, everything! You are part of an unbroken lineage starting from the very first life-forms on a planet that only exists because certain physical laws allow it. So we are here because the Universe enabled it, which means we are assholes because the Universe made us this way. But the real question is how and why it decided we should be such bastards.

Our ability to understand and manipulate the rules of the

Universe has alleviated many of the environmental pressures that our ancestors would have faced. From building houses and skyscrapers to programming robots to have sex with us when no other human will. We have indeed achieved some amazing feats, yet our skulls still house essentially the same cave-dwelling simian brain of our predecessors. This means it is just as easy to fool each other and also ourselves as it was for our troglodyte forefathers.

Let's start with *moral licensing,* a subconscious thought process that weighs up the previous good and bad things that you have done when you are faced with a right or wrong choice. Imagine walking your dog and they take a shit in a neighbor's garden bed. It's off the path, so no one will step in it, but you know Mary loves those azaleas and will most certainly come across either a chocolate or vanilla swirl depending on when she next trims her front garden. But then you remember the time you brought the trash out for her, and that seems like a fair deal for your "bend and pretend." This is an example of moral licensing—it is acceptable in your monkey brain to do something bad because you already did something virtuous. In fact, studies have shown that moral licensing can also act prospectively—meaning you can act immorally because you know that sometime in the future, you will do something good. In short, you chose the right chapter of this book to read today, asshole.

Moral licensing is just one type of unintentional subconscious mental gymnastics that impacts many aspects of our lives. We love to blame people...scratch that, we *need* to blame people when things go wrong. But maybe we really ought to be blaming the Universe, which—through the mechanics of evolution—has allowed us to develop some neat cognitive tricks to ensure that even when things turn to shit, we can still feel better about ourselves in light of making the wrong decision.

This chapter isn't intended to take anything away from the complexity of our brains. They are the crown jewel of evolution. Life has spent around 3.5 billion years evolving on a planet that only came to be around 4.5 billion years ago. We have gone from single-celled organisms with the most basic of functions to sending rockets to the moon and creating sexy robots. If you don't think that's progress, then you clearly haven't masturbated using a machine.

Charles Darwin—who famously put forth the theory of evolution by means of natural selection—was fascinated with how our own social and personal behaviors could have evolved. Some scientists will argue that all of our psychological mechanisms are rooted in evolution. Darwin's work on our social and personal behaviors did not have the same impact as some of his earlier projects. But in fairness, when one of those projects is in the running for the single greatest ideas any human mind has ever developed, any subsequent work is going to be underwhelming by comparison—kind of like being the fifth Baldwin brother. Or was he the fourth? Ah, who cares. But just like the Baldwin brothers, this idea would spark much controversy, and although no photographers were punched in the face, there still to this day remains a debate over just how much influence our environment and genetics play in our psychology.

The evolution of our brain from whatever primitive thing it was to a soft mush of pruny flesh is certainly less exciting than the venom of a snake or the teeth of a shark. But when you consider how our everyday behaviors are rooted in evolution, it becomes fascinating to think about our ancestors. Going back hundreds of thousands of years, how did they express emotions, grieve, attract a mate, or maybe steal all the soft leaves for themselves so the others in the cave couldn't wipe their ass during lockdowns?

Living in this universe is like making a deal with the devil. We get to experience love, happiness, joy, empathy, and a host of other feelings that give us a tingly warm feeling in the cockles of the heart. But these emotions often stir up a million other things in our head alongside the realization that we aren't really in that much control over what comes and goes in our brain. But let's not be too quick to dismiss certain behavior as a consequence of the unconscious mind—we can be conscious assholes too.

There are many times in our life when we are happy for people to think less of us. How many times have you thought or said *I don't give a fuck what Alec Baldwin thinks?* It must be once or twice a day. Consider the example of the soft leaf toilet paper shortages in many countries during the COVID-19 pandemic. Many people didn't care if others couldn't wipe, as long as they could get their hands on ten 24-packs of 3-ply ultrasoft toilet paper—despite having several rolls at home already and being told repeatedly that there was plenty to go around... or front to back. This type of thinking, known as *zero risk bias*, was probably essential for our ancestors' survival during times of scarcity. How useful this thought process is today is questionable. But we also have the luxury of clean butts to enjoy while being able to explore these questions.

Essentially, zero risk bias describes how people, during

times of elevated risks or dangers beyond their control, try to completely eliminate other, less severe risks or dangers within their control. This sounds fine at face value, but often this thinking actually exposes us to more risk, and we would have been better off reducing the larger risk rather than eliminating the small one. So what has this got to do with you having to wipe with a sock that one time during a lockdown? Well, toilet paper is cheap, and even though it won't directly prevent the spread of COVID-19, it is an easy action to take with very little risk involved. Having a stockpile of toilet paper will give you a sense of control—as well as amazing fort-building capabilities—and will ensure that you have eliminated at least one threat: a shortage of things to wipe your ass with.

Although hilarious in this context—unless you're the one wiping with a crusty gym sock—it does highlight a rather concerning flaw in the way we approach risk management when looking at bigger, more serious issues. Zero risk bias can also affect decisions that can ultimately impact the health and safety of others and all because feeling in control of a situation—regardless of whether these feelings are based in reality or relevancy—helps us cope with the uncertainty and chaos of the Universe around us.

It isn't all bad though—there are upsides to self-serving jackassery. Some research suggests being an asshole in the office actually has some benefits, not just for you, but also your colleagues. Sounds too good to be true, right? Well, it's not. Gossiping and ostracism can actually have positive effects in reforming bullies and avoiding nice people being taken advantage of, thus encouraging group cooperation. Gossiping—or "talking shit" as it is affectionately known—about others can rationalize another's behavior and allow the alignment of values whilst discouraging people from taking similar actions that would otherwise lead to group instability in the future. So don't feel bad about that time you were bitching about Bryan at work for stealing toilet paper from the men's bathroom—classic Bryan. The more people you tell, the less likely they are to be a hoarding thief with wanton disregard for workplace hygiene.

Us humans are the product of billions of years of evolution that has forced this behavior on us. We don't *want* to do these things. If it were up to us, we would probably want what every middle-class suburban home has written on some cheap ornament—we want to *live, laugh, and love,* or some shit. But this isn't how the Universe works. So deal with it and be proud that your ancestors fucking rocked it! If that meant being a complete asshole to ensure more food, protection, and sex—thus passing their genes on to future generations of assholes who would grow up to do the same—then so be it; it's what the Universe wanted.

The Universe has ingrained in us a genetic toolkit for survival, and one of these tools is the need to be an asshole. Sure, some of us default to it more often and sooner than others. But whether you like it or not, this appears to be a very successful characteristic when it comes to managing life... or pushing an old lady out of the way for some 3-ply.

Reason 3

Radicalized oxygen is trying to kill you

Focus and take a deep breath. Close your eyes... Wait, don't do that. Hands where we can see them. Let's start again...

Focus and take a deep breath. Breathe in...and now breathe out. Breathe in again... No, not that much! Are you mad? You know what? Just stop. No, don't stop breathing! That's just reckless. Look, just try to get through this chapter before you pass out. Just make sure you're breathing exactly the right amount. Perhaps it's best not to think about it too hard.

Oxygen was first discovered in 1774 by Joseph Priestley—who had a knack for getting high off his own supply. Another of his discoveries, nitrous oxide, or as it is more commonly known, laughing gas, would've been rather enjoyable for him. For obvious reasons, he probably spent more time experimenting with nitrous oxide than oxygen. But, unbeknownst to him, he had discovered a molecule so essential to our very existence that even when he was high as a kite, Priestley could not have imagined its importance. Oh, and just to be clear, we're talking about oxygen.

When we think of oxygen, we usually look at it in the context of necessity. Something that, if not consumed within a few minutes, can assure certain death. This life-giving molecule is also associated with health, not only for us, but also many facets of our environment. Around 300 million years ago, there was 10 percent more oxygen in the Earth's atmosphere. Consequently, insects were significantly larger because their small respiratory tracts were more than adequate to deliver enough oxygen to the rest of their body making them more sizable than the ones we see today. In fact, fossil evidence has shown that a dragonfly's wingspan would measure the same distance as laying 10,034 three-toed sloth sperm head to tail, or about 70 cm.— breathtaking stuff.

At a fundamental level we, and pretty much every other earth-dweller, undergo a process known as respiration whereby we inhale oxygen and exhale carbon dioxide. But have you ever asked yourself why you need oxygen? No, you haven't. It's okay—we didn't either until we got a book deal to write about it. Anyway, the answer is apparently quite simple. At the cellular level, we need oxygen to make energy, which is what we need to do important things like sit slouched over thumbing our way through Facebook shitposts.

Just like a lit candle, if your body is not constantly supplied with oxygen, then your flame will go out. That is to say, you'll die. In the most basic terms, oxygen combines with the sugar in an unnecessarily complicated and boring series of biochemical reactions to make something called adenosine triphosphate, or ATP. ATP is the heat that keeps your candle burning. If you woke up in high school biology at just the right moment, you were probably startled by your teacher enthusiastically referring to the *mitochondria* as "the powerhouse of the cell." Even if you weren't paying attention, internet memes have probably caught you up. The mitochondria are where most of our ATP is synthesized, making them mini power stations. To be clear, for our more youthful readers, "powerhouse" used to be a synonym for power station—the mitochondria is not supposed to be analogous to Arnold Schwarzenegger back when he had a reduced sperm count.

There is no question that oxygen is crucial to our survival. Most people can only go about three minutes without it before serious irreparable complications set in. But the Universe rearranges oxygen in multiple ways as if it was a Lego piece, and anyone who has stood on a Lego barefoot knows how vicious it can be.

To fully grasp how and why oxygen is trying to kill us, we first must understand some pretty simple chemistry. Whoa! Whoa! Just wait a minute. Before you get all "but I'm not good at sciencing" and flip to the next chapter, just hear us out for a moment. It has got to do with a group of chemicals called free radicals. Doesn't that sound dope?

There are many types of radicals, but the most harmful ones all seem to contain oxygen—hence our hostility toward it. Oxygen is an electron thief and likes to hoard electrons that don't belong to it. This leads to molecules that are unbalanced, surrounded by electrons yearning for lost partners. You see, electrons (the negatively charged particles that orbit an atom) like to be paired up with other electrons. These oxygen-based free radicals become very unstable as the unpaired electrons frantically search for a new partner; much like a recently divorced forty-five-year-old in a trendy nightclub, they will react with whatever they rub up against. Inside living things, this could include cell membranes, proteins, and DNA. Essentially the radical has the potential to bump into, and damage, a bunch of stuff you wish it would stay away from.

Interestingly, your body actually uses free radicals to help fight off pathogens in a process known as phagocytosis, which involves the cells in your immune system "eating" harmful foreign entities. Your phagocytic cells, through a number of chemical reactions, create free radicals that attack and break down the same parts of the invading cells that we were worried about protecting two paragraphs earlier. How's that for a twist? It's almost like oxygen was a story written by M. Night Shyamalan...except, the story of oxygen has actually won some awards, so maybe not.

There's more good news, too. Readily available chemicals that oppose these free radicals can be found all over and even made in your body. These chemicals are called *antioxidants*, and people who frequently exercise and maintain a healthy diet have an ample supply of these that enable them to handle any naturally occurring free radicals. For the rest of us, here comes the bad news. Things like smoking, drinking, indulging your appetite, and going out in the sun too much—in other words, having fun—can cause a phenomenon known as *oxidative stress*. This is where our bodies become overwhelmed by free radicals and as such, are unable to get rid of them. This inundation of free radicals plays a major part in many chronic degenerative health issues like cancer, autoimmune diseases, aging, and Alzheimer's disease, to name just a few.

You're probably thinking *ahh, yes, but what about the antioxidants in my all-natural grass-fed organic goat's-milk yogurt?*

If only our body could balance a free radical bombardment from a long weekend at a festival with an antioxidant super dose. Then we could just suck a goji berry and kale smoothie down before getting all Benjamin Button on everybody's ass. But we're sorry, the Universe has that stopgap solution already figured out.

Like so many fad diets, megadosing on antioxidants is not the answer. In fact, overconsumption of antioxidants has been linked to an increased risk of cancer and is even known to have a pro-oxidant effect. So don't start thinking that because something is marketed as high in antioxidants it is somehow better for you. It is no more nutritional than normal fruit and vegetables. In short, superfoods are a load of bullshit that trick people into thinking that eating healthy has to be super expensive. It's not. It's actually only moderately expensive, so that only leaves out about one third of the world.

Anyway, what has all of this got to do with the Universe? Well, we don't really have a choice in how life evolves, nor do we have any say in what can sustain it. Why then would the Universe turn our existence into such a fine balancing act? Let's take a moment to recreate an event that, unfortunately for humanity, many see as more plausible than Darwinian evolution—a conversation between Adam and God.

"Here, you'll need this," God says.

"What is it?" replies Adam.

"This? Oh, it's oxygen. Ensure you breathe in 1.5 liters of it every minute."

"Wow. That's a lot," Adam says with a confused look on his face.

"Yeah, but if you don't take in enough, it could cause a slight problem."

"What slight problem is that?" Adam asks.

God whispers under her breath, "Umm...death."

"Death!?" Adam responds in a somewhat harsher tone.

"Okay, look, it's a bit of an oversight on my part."

"A bit of an oversight? I'd say it's an enormous oversight."

"You think that's bad, wait till I have to chat with Eve about childbirth," God mutters out of the side of her mouth as she points her thumb toward an unsuspecting Eve. "Okay, Adam, I need you to take a big breath for me. Big breath. Come on, who's a good boy?"

Adam breathes in.

"No, that's not enough. You're turning blue. Oh no, you're going to pass out!"

Adam, panicking, starts breathing more heavily.

"No. That's too much. Stop. Oh, no..."

"What's wrong?" Adam worryingly asks.

"You've just given yourself a genetic disease," God says as she turns and mutters to herself. "Piece of junk..."

"What was that?" Adam assertively questions.

"Nothing." God replies "Here, have an apple. You'll need the antioxidants."

Acknowledgments

There are so many people to anti-thank here, but now is not the time. First and foremost, though, I need to thank Lindsay. Thank you for always laughing at my jokes and scowling at the ones that weren't funny. Thank you to my children (whom I didn't read this book to...yet) for making me laugh and laughing with me. I owe my generally optimistic view of humanity to my dad and his dad before him—thanks for the cat, Grandpa. Last but far from least, Mom, thank you for being my biggest fan.

I also want to thank some early readers who provided suggestions and encouragement (but whom I can't claim endorse anything I've written here!): Sarah Kaiser, Wade Fairclough, Byrne Laginestra, and Ryan Ferrie. A special thanks goes to my editor, Anna Michels, who challenged me to be clearer and funnier and to use fewer f-bombs.

About the Author

Chris Ferrie is an associate professor at the University of Technology Sydney in Australia, where he researches and lectures on quantum physics, computation, and engineering. He is the author of more than fifty children's books about science, including *Quantum Physics for Babies* and *There Was a Black Hole That Swallowed the Universe.*